ENVIRONMENT AND LIVESTOCK IN INDIA

With a Comparative Study of the Indian and US Dairy Systems

ENVIRONMENT AND LIVESTOCK IN INDIA

With a Comparative Study of the Indian and US Dairy Systems

S.N. MISHRA
A.K. DIKSHIT

MANOHAR
2004

First published 2004

ISBN 81-7304-563-1

Published by

Ajay Kumar Jain for
Manohar Publishers & Distributors
4753/23 Ansari Road, Daryaganj
New Delhi 110 002

Typeset at

Digigrafics
New Delhi 110 049

Printed at

Lordson Publishers Pvt. Ltd.
Delhi 110 007

Contents

Preface

At the Kyoto Protocol convention on greenhouse gases emission reduction in 1997, some of the developed countries with the United States in the lead refused to accept any binding condition unless major developing countries, India and China in particular, also became a party to the binding agreement. Being resident in Tokyo and teaching at the prestigious Hosei University, media persons quickly sought me out for reaction. My usual reaction was: well, it is they who have created the problem by accumulating emission over the last two and half centuries. But then I added that the poor people in the developing countries, whose livelihood depends on natural resources, positively contribute to environment protection instead of damaging it. Look at millions of poor Indian families, which carry on their agriculture using animal energy instead of fossil-fuel energy. Don't they prevent carbon dioxide emission thereby? If one was to work out carbon dioxide emission per unit of food produced, India's emission would be very much less compared to that of a developed country.

Back in India in 1998, I sent out a short three-page proposal on environmental contribution of working animals stock in India to Dr Amrita Patel, Chairperson of the National Dairy Development Board (NDDB), to explore whether she would be interested in supporting research on the subject. Not only was her response positive, she also asked me to enlarge the scope and include milch animals too in the proposal, which I did. The final outcome is the study presented here. It includes not only the environmental contributions of the working and milch animals but of the total bovine population of India. Although in my professional life I have done considerable amount of research on India's livestock economy, its planning and development, I had little idea of how much work this new environmental dimension will involve. As it happens, it has involved considerable effort. Yet this study is not complete. Small ruminants, sheep and goat, are out of its scope. Similarly, my original idea of making a comparison of greenhouse gases emission per unit of food grains produced in India with that of a developed country remains to be put into effect. Still, I believe that the task we have accomplished in this study is not small. The study, I hope, will be useful to students of environment, to concerned scientists, to policy makers, and in international negotiations on greenhouse gases emission reduction.

If there is one person without whose support this study would not have been possible, it is Dr Amrita Patel, Chairperson of the NDDB. In making the grant for the project she asked us no questions about the Society for Economic and Social Research (SESR), for instance, what sort of institution it was or whether it had the necessary infrastructure to carry out the project. Few persons in positions of power and authority dare take decisions on trust. At the SESR, we have made our best effort to fulfil her expectations. I would like to record my immense personal gratitude to her. The other person whose support is incomparable is Shri Jaswant Rai Arora, proprietor HCMR Farms in the Mandoli Village of Delhi. Free office premises of the SESR and the amenities are due to his sense of philanthropy for good causes. I wish to express my deep sense of gratitude to him on behalf of the SESR, and my own behalf. My friend, Surendra K. Kaushik, Professor at the Pace University, New York and Associate Director of its Business School has been a constant source of support and inspiration since the formation of the SESR. The SESR runs a centre in his name, the Kaushik Centre of Contemporary Issues (KCCI), which he has endowed. I wish to convey my grateful thanks to Dr B.K. Ganguly, the NDDB Coordinator of the project, for his unflinching support throughout the course of this study and until its publication. Finally, I am very grateful to NDDB for supporting this study and its publication. The NDDB, however, is not at all responsible for the contents, or the views and opinions held in this book; the responsibility rests solely with the authors.

Chairman S.N. MISHRA
Society for Economic and Social Research
Village Mandoli, Delhi
September 2003

Acknowledgements

This study was made possible with support coming from many quarters. We are deeply indebted to the National Dairy Development Board (NDDB) for financial support. Dr B.K. Ganguly, Coordinator of the project from the NDDB side, was extremely considerate and helpful. The animal scientists, statisticians, and economists who participated in the workshop held in January 2000 to discuss the environmental model of the livestock production system (Chapter 4) deserve our thanks for their helpful comments and suggestions. In particular we would like to mention Dr Prem Narain, former Director of the Indian Agricultural Statistics Research Institute (IASRI) and an eminent man of animal science. He carefully read through the model and gave us the benefit of his comments and suggestions. Dr Shivtar Singh, with his long experience of research in livestock statistics, was always available for checking statistics, and for general consultation. We would like to express our gratitude to Mr Allen J. Baker of the Economic Research Service, United States Department of Agriculture for help in accessing US livestock related data on the Internet and for providing information on certain parameters which enabled us to make a comparison of the US and Indian milk production systems for their environmental effects. This study in draft form was presented at a seminar held on 3 October 2001 at the NDDB, Anand (Gujarat). Many comments and suggestions came up during the discussion. Some written comments too were offered by the NDDB colleagues, who had read the draft. Thanks are due to the seminar participants and those NDDB colleagues who offered comments and suggestions on reading the draft.

We are extremely grateful to Dr S.K. Upadhyay who red through the manuscript before it was sent to the press. Being a mathematical statistician he meticulously checked the model in Chapter 4 and statistical tables in the text. The 'Glossay of Notations Used' was also prepared by Dr Upadhyay for the reader's convenience.

Dr K. Lal and Dr P.D. Sharma have been a source of optimism in times of pessimism. We received valuable support from the library staff of the Institute of Economic Growth, Delhi, in particular from Shivdasan and the Assistant Librarian, P.K. Jain. T.C. Sharma, Y.V. Chandan, and Sushil K. Sen helped in many ways. S.P. Sharma provided statistical assistance and did the job of information gathering. Word processing was initially done by Satya Prakash, and in the final stages by Ravinder Singh. The Society for Economic and Social Research (SESR) being located in a village, access to libraries and offices in the city was made possible by Jai Singh. We are thankful to them all.

We would like to express our gratitude to NDDB for the grant, which made possible the research work, and financial support for its publication. The NDDB, however, bears no responsibility for the findings reported or the views and opinions held here. The responsibility is entirely our own. Finally, Mr. B.N. Varma, editor of the book, deserves our thanks for making the text readable.

SESR, Village Mandoli, Delhi S.N. MISHRA AND A.K DIKSHIT
September 2003

Glossary of Notations Used

Symbols	*Description*
f_{jy}	j-th type of feed quantities fed to a young cattle
d_j	Dry matter fraction of j-th feed
e_j	Energy value per unit of j-th feed
gei_y	Gross energy intake of an young cattle
CH_{4y}	Methane emission per young cattle per year
CH_{4yw}	Methane emission per year per young cattle in weight (kg)
TE_y	Total methane emission per year in weight from young cattle
Y	Population of young cattle
F_i	Population of adult females in i-th category
CH_{4c_iw}	Methane conversion factor of i-th category adult females
TE_c	Methane emission from total adult females
n	Number of categories of adult females
CH_{4m_iw}	Methane conversion factor for adult males in i-th category
M_i	Population of adult males in i-th category
TE_m	Methane emission from male cattle stock
n'	Number of categories of adult males
CTE	Total methane emission from cattle stock (C for cattle stock)
d^*_y	Per day dung output per young cattle
α	Dry fraction of dung
β	Fraction of ash content in the dry dung
VS_y	Volatile solids in dung output per day per young cattle on dry matter basis net of ash content

F_y	Feed intake per young cattle per day on dry matter basis
r	Feed digestibility ratio
F'_y	Dung on dry matter basis evacuated per young cattle per day
θ	Cubic meter CH_4 per kg of VS (maximum potential)
ϕ	Realization fraction (or methane conversion factor)
CH_{4ysw}	CH_4 emission per year per young cattle in weight
TE_{ys}	Total CH_4 emission per year from total young cattle stock
CH_{4c_isw}	CH_4 emission (weight per year per adult female cattle in i-th category)
TE_{cs}	Total methane emission (in weight) per year from dung output from total cattle female stock
CH_{4m_isw}	Methane emission (in weight) per year per adult male cattle in i-th category
TE_{ms}	Total methane emission (in weight) per year from dung output of total adult males stock
CTE_s	Total methane emission (in weight) per year from dung output of total cattle stock
CTE_{CH_4}	Total methane emission (in weight) per year from both enteric fermentation and manure management of the total cattle stock
μ	Fraction of dung output used as manure
N	Nitrogen
P_2O_5	Phosphoric acid
K_2O	Potassium oxide
χ'	Ammonium sulphate replaced by dung per animal per day
χ	Ammonium sulphate replaced by dung from total young cattle population per year
χ_{iy}	Replacement amount of the i-th chemical fertilizer from the dung output from total young cattle stock
π_{id}	Fraction of i-th nutrient in the wet dung
π_i	Fraction of i-th nutrient in the i-th fertilizer

χ_{ij}	Replacement amount of i-th fertilizer from j-th category of cattle (male and female)
d^*_j	Per day dung output per animal of j-th category of animal
C_j	j-th category of cattle population
x_i	i-th category of fertilizer saved due to use of dung manure of the total cattle stock
n''	Number of categories of cattle (male and female)
e_{iCO_2}	Carbon dioxide emission (volume) per unit of i-th fertilizer manufacture
E_{iCO_2}	Total carbon dioxide emission(volume) saved from i-th fertilizer due to its replacement by manure
E_{CO_2}	Total carbon dioxide emission (volume) saved from all fertilizers due to its replacement by manure
ρ	Ratio of global warming potential (GWP) of methane to carbon dioxide
E_{CH_4}	Total methane equivalent of carbon dioxide from use of dung manure
D_c	Dung cake output per year from the cattle stock
p	Carbon fraction of dung cake
p_0	Fraction of carbon oxidized
C_f	Carbon released from burning D_c
e_{CH_4}	Emission ratio of CH_4 to carbon dioxide in the process of burning
D_{cCH_4}	Total methane released from burning D_c
a_t, a_{t+n}	Number of working animals (per unit of net sown area) at time t and t+n
b_t, b_{t+n}	Number of tractors (per unit of net sown area) at time t and t+n
s	Economic rate of substitution between working animals and tractors
A	Number of working animals
T	Number of tractors required to replace A
p^*	Diesel used per tractor per year
P	Total quantity of diesel required to run T number of tractors
σ	CO_2 emission (in weight) per unit of diesel used

$\overline{CO_2}$	Total carbon dioxide emission from diesel used in tractors
E'_{CH_4}	CH_4 equivalent to $\overline{CO_2}$
CCH'_4	Total CH_4 emission due to cattle stock
CCH''_4	CH_4 emission prevented due to use of animal energy in agriculture and use of dung as manure
CCH_4	Balance of CH_4 emission from the cattle stock
e	Breeding efficiency (ratio of cows giving birth to a calf (cows in-milk) to total breedable cows)
k	Inter-calving period, measured in calendar year
q	Methane emission per cow in-milk per year from enteric fermentation
q′	Methane emission per cow in-milk from manure management
C_m	Population of cows in-milk
CH_{4m}	Methane emission from cows in-milk
g	Methane emission per dry cow per year from enteric fermentation
g′	Methane emission per dry cow from manure management
C_d	Population of dry cows
CH_{4d}	Methane emission from dry cows
CH_{4b}	Methane emission from total breedable cows
O_c	Total output of cow milk
V	Emission per unit of milk considering only cows in-milk
V′	Emission per unit of milk considering total breedable population
x_1	Energy intake per day per young cattle from concentrates
x_2	Energy intake per day per young cattle from dry fodder
gei^*_y	Gross energy intake per day per young cattle from by-products
Z_y	Quantity of green fodder (dry matter basis) required to replace gei^*_y
e_3	Energy (M cal) per unit of Z_y

z'_y	Green fodder (on dry matter basis) required to replace energy per day per young cattle from by-products
z''_y	Green fodder required per year to replace energy from by-products
λ	Yield of green fodder per hectare
l_y	Area required to produce z''_y
L_y	Total area required to produce green fodder to replace energy from by-products
gei_{c_jp}	Energy intake from by-products per day per animal of j-th category
z''_{ij}	Green fodder per year per animal of j-th category required to replace energy from by-products
l_{c_j}	Land area required to produce z''_{c_j}
L_{c_j}	Annual requirement of area for producing green fodder to replace energy intake from by-products for total cattle in j-th category
L_{gc}	Total land per year required to produce green fodder to replace energy obtained from by-products for total cattle stock
ε	Rate of substitution of dry fuel wood for dung cake (in terms of thermal energy)
W_t	Total quantity of dry fuel wood required to replace D_c
q^*	Yield per ha of fuel wood on dry matter basis
L_t	Total land area required to produce W_t
h	Gestation lag (in years) between planting and harvesting of trees
L'_t	Total land area required to replace Dc under assumption of gestation lag between planting and harvesting of wood
MT	Million tonnes

STANDARD EQUIVALENTS

1 tonne of oil equivalent (toe)	1×10^{10} calories
10^3 toe	41.868 TJ
1 short ton	0.972 tonne
1 tonne	1.1023 short tons
1 tonne	1 megagram
1 kilotonne	1 gigagram
1 megatonne	1 teragram
1 kilogram	2.2046 lbs
1 hectare	10^4 m^2
1 calorie IT	4.1868 Joules
1 atmosphere	101.325 kPa

Source: IPCC (1995).

CONVERSION FACTORS

1 metric tonne = 2204.622 pounds
1 kilogram = 2.2046 pounds
1 acre = 0.4047 hectare
1 square mile = 640 acres = 259 hectares
1 gallon = 3.7853 litres

US MEASURES EQUIVALENT TO METRIC MEASURES

S. No.	Commodity	Unit	Approximate net weight	
			US (pounds)	Metric (kilogram)
1	Corn: ear, husked	Bushel	70	31.8
2	Sorghum	Bushel	56	25.4
3	Wheat	Bushel	60	27.2
4	Barley	Bushel	48	21.8
5	Cotton	Bales (gross)	500	227
		Bales (net)	480	218
6	Soybean	Bushel	60	27.2

Source: USDA, *Agriculture Statistics*, 1994, NASS.

1. Introduction

1.1 GENERAL BACKGROUND

The relationships between livestock population, livestock production system, and environment are fairly extensive and deep. Livestock, being a natural resource, are themselves a part of environment. Environment has a determining influence on several of the characteristics of a livestock population—its biodiversity, its reproduction parameters, and productive traits, for instance, notwithstanding attempts to change those characteristics through anthropogenic interventions in natural breeding and feeding systems. Seen in terms of its territorial size and spread across latitudes and longitudes, its physiography, soil-structure, topography, climatic conditions, and micro-ecosystems, India's environment is quite diverse. The climate alone varies from Tundra-like conditions in the Trans-Himalayan region to near tropical conditions in the peninsular south. This diversity has a bearing on the biodiversity (breed-types) of different species of livestock in the country, animals' body-mass, and their reproductive characteristics.

The livestock population draws or receives energy from the environment in the form of feed from grazing or browsing, from by-products of agriculture, and from cultivated green fodder. It returns the energy in part to the environment and in part to the society. Animal droppings on grazing lands fertilize the soil. Dung manure performs the same function. The energy derived from working animals still remains the mainstay of Indian agriculture, despite growing tractorization. Society receives part of the energy in the form of consumption goods, milk and meat, not to mention a large number of residual products from dead and slaughtered animals.

In the process of conversion of feed into nutrients by ruminant livestock, a part of the energy from feed is used up in methane formation, which the animals release to the atmosphere largely by eructation. This is termed as emission from 'enteric fermentation'. Methane is formed in the digestive tract of the animal during the process of feed fermentation. There is also some emission from dung, depending upon whether dung is managed in aerobic or anaerobic conditions. Methane emission from livestock sources of a country constitutes a part of the global methane emission, which has an environmental impact. Together with other greenhouse gases it is expected to cause global warming and climate change a century or so hence, if the emission of methane and other greenhouse gases continues at the current rates. Ruminant livestock and rice-fields in that order are supposed to be the top ranking sources of methane emission worldwide. Incidentally, the predicted climate change will not be a disaster for every country around the globe. Futuristic predictions about the effects of climate change show that countries in the temperate regions are likely to benefit from it (Cooper 2001).

India has the world's largest livestock holding, accounting for about 16 per cent of the cattle and more than half of the buffalo population of the world (GOI 1999). The livestock population base of the production system is accordingly large, though the productivity of the system may be comparatively low. By virtue of the sheer size of her ruminant livestock population, India is expected to produce more methane than any other country. The same should be true of other environmental dimensions of India's livestock production system. For instance, perhaps no other country in the world provides renewable and sustainable energy to its agriculture from a working animals' stock as large as over 70 million. Similarly, perhaps no other country in the world maintains its livestock population mostly on by-products of agricultural production, some concentrates, and more of crop residues requiring no additional land for their production, as India does.

India's livestock production system is equally important from the economic angle. Milk is the chief source of animal protein for Indians. The per capita per day milk availability, however, is still much below the nutritional norm recommended for an average Indian. The nutritional norm of course varies according to age groups and between the vegetarian and non-vegetarian sections of the population. But taking them all together the norm works out to about 250 gm of milk per day per capita (Indian Council of Medical Research 1971). Availability, on the other hand, was no more than 182 gm in 1992–3 (GOI 1999). None the less, the relative position of the livestock sector in the national economy has been improving. Currently it accounts for about a quarter of the value of output from agriculture and allied activities, and about 6 per cent of the nation's gross domestic product (CSO 1996). Tens of millions of rural families, specially from the poorer sections, depend on livestock as a source of livelihood.

1.2 Environmental Concerns

In the post-independence period, though a good deal of research on various aspects of the livestock economy and its development has been undertaken (Mishra 1995), not much attention has been given to the related environmental aspects. W. Burns was perhaps the first to recognize early on the role of environment in productivity of livestock in India. He pointed out the existence of an inverse relationship between milk yields of cows and buffaloes and the intensity of rainfall, noting that as one moved from the relatively dry regions of the north-west to the wet regions of the east and south-east, milk yield of the animals was seen to decline (Burns 1944). Later on, Mishra formulated a regional livestock planning model for India in which the country was divided into six livestock regions based upon intensity of rainfall, physiography, and territorial contiguity (Mishra 1978). Both Burns and Mishra were moved by the fact that India's environmental diversity made a difference to animals' productivity, and implicitly recognized the difference it made to the reproductive traits of the livestock population.

The feedbacks from the livestock population to the environment came under public discussion for the first time during the 1980s. The background to it was the rising environmentalist movement across the world following the first international conference on environment and development held at Stockholm in 1972. Domestic environmentalists made a claim that the small ruminants, sheep and goat, were damaging the environment in the fragile zones of the country. The Government of India appointed an expert group to study the problem and suggest protection measures. The study came to the conclusion that the environmentalists' claim was unfounded (GOI 1987).

Another feedback to global environment, namely emission of methane from livestock sources of different countries, apart from emission of other greenhouse gases and sources, became a subject of international importance during the late 1980s, in the background to the forthcoming Earth Summit at Rio de Janeiro in 1992.

In the Kyoto Protocol Convention on the reduction of greenhouse gases in 1997, some developed countries with the United States in the lead made a demand that large developing countries like India and China must also be made a party to any binding international agreement on greenhouse gases reduction. Even though the US administration which took charge in January 2001 has already rejected the voluntary commitment the previous administration had made at Kyoto, such demands will continue to be made in future. Accordingly, quantitative estimates of greenhouse gases emission in the developing countries, including from livestock sources, will continue to be made in the developed world. It is not surprising, therefore, that the first estimates of methane emission from India's livestock sources were made at the US Environment Protection Agency (Ahuja 1990) and by the World Resources Institute (WRI 1990). The estimates were for 1990 and close to each other—10.4 million tonnes by Ahuja and 10.0 million tonnes by the WRI. Since then the WRI has routinely been making estimates of methane emission from livestock sources in India, and releasing these estimates in their annual report. The latest available figure is 12 million tonnes for 1991.

Based on results of experiments carried out at the National Dairy Research Institute (NDRI), Karnal, G.P. Singh reported a figure of 9.023 million tonnes for methane emission from livestock in India in 1993, nearly 1 million tonnes less than the earlier estimates for 1990 (Singh 1997). Interestingly, another estimate of methane emission made at the Centre on Global Change, National Physical Laboratory, Delhi worked out to 8.468 million tonnes in 1992, close to 2 million tonnes less than the earlier estimates for 1990 (Bhattacharya and Mitra 1998). It is also worthwhile to note that using the Intergovernmental Panel on Climate Change (IPCC) 'default rates' (methane conversion factor per animal per year) recommended for South Asia, Bhattacharya and Mitra have reported methane emission from enteric fermentation alone to be as high as 10.8308 million tonnes in 1992, compared to the figure of 7.563 million tonnes (included in the total of 8.468 million tonnes) for the same year based on their own methane conversion factors, there being, thus, a difference of more than 3 million tonnes for the same year. G.P. Singh's estimate for 1993 is also derived from enteric fermentation alone. However, his livestock population base is close to the livestock census population of 1987, rather than the population in 1993 as he has claimed. That being the case, and supposing that his estimate of methane emission (9.023 million tonnes) relates to the year 1987, then it is higher by about 1.5 million tonnes compared to that of Bhattacharya and Mitra (7.563 million tonnes) for 1992.

Our purpose here is not to make a critical appraisal of the studies on methane emission from livestock sources in India. While all these studies relate to livestock populations within a short span of 5 years, yet there is much difference between their estimates of methane emission from India's livestock sources. The inevitable conclusion is that a great deal of uncertainty surrounds the estimates, and none can be taken as definitive.

Having said that, it is important to recognize that scientific attempts to estimate methane emission from India's livestock sources, and earlier on, environmentalists' concern about the threat to environmental damage in the fragile ecozones of the country from free livestock grazing were the first attempts at exploring the feedback from the livestock population to the environment. Both of these, however, are negative feedbacks. The positive feedbacks, for instance, those arising out of the energy exchange between the livestock production system and the environment, have not yet been explored.

1.3 The Scope of the Study

This study makes an attempt to provide a more comprehensive examination of the relationship between the livestock production system and environment. Not just the negative feedback in the form of methane emission but also the positive feedback, such as prevention of carbon dioxide emission due to animal energy used in agriculture in place of fossil fuel, or saving of the natural resource, land, due to agricultural by-products used as animal feed, have been included in the study. The study is focused on cattle and buffaloes, the bovine stock which accounts for 90 per cent of the country's livestock population in terms of feed consumption, in other words, gross energy intake from the environment. It has not been possible to include small ruminants mainly because national level data on their feed consumption rates, including the biomass grazed per animal per day are not available. The studies on methane emission mentioned in the preceding section are all based on feed *requirement* rather than on feed *consumption*. Requirement estimates are problematic since they are based on assumed average body weight of animals at the all-India level, and exogenously given requirement *norms* for maintenance and levels of production, for instance milk yield of a cow in-milk. We have preferred to avoid this course. The study has been carried out at the all-India level. Now that the agro-ecological regions of the country have been identified (National Bureau of Soil Survey and Land Use Planning 1992) a more realistic study at the level of these regions incorporating environmental diversity should become possible in future.

1.4 Organization of the Study

Chapters 2 and 3 are preparatory to understanding of the environmental model presented in Chapter 4. Chapter 4 requires some knowledge of mathematics on the part of the reader. The general reader can skip it without any loss, and go to Chapter 5 onwards which presents the computational results of the model. Chapter 2 provides a descriptive account of the interrelations between livestock and environment in the specific situation of India. The livestock production system of the country has been described in Chapter 3. The computational results on methane emission from enteric fermentation, 'manure management', and from burning of dung cakes are presented in Chapter 5. Chapter 6 is devoted to carbon dioxide emission prevented due to animal energy used in agriculture. Land saving due to the livestock production system arising from agricultural by-products feed, and dung cake used as domestic fuel in place of fuel wood, is the subject of discussion in Chapter 7. Chapter 8 is a short note on the estimate of chemical fertilizers saving due to dung used as manure. In Chapter 9 an attempt is made to value the positive environmental effects of the livestock production system using the framework of economic cost-benefit analysis. Chapter 10 is devoted to a comparison of the environmental effects of India's milk production system with that of a developed country, the United States of America. Rather than merely compare methane emission from livestock sources, a comprehensive comparison, taking into account environmental effects both on the input and output side of the milk production system of the two countries, has been attempted in this chapter. The last chapter provides the summary and conclusions of the study.

2. Environment and Livestock

2.1 Introduction

Research in the economics of livestock in India has hardly touched upon the issues relating to environment and the livestock production system of the country, though there is an interactive relationship between the two. Environment has a determining influence on the biological characteristics of the livestock and their productivity. As environment varies, so do the bio-characteristics of any given species of animals and their productivity. W. Burns was perhaps the first to recognize at least the influence of climatic conditions on cow milk-yield. Taking rainfall intensity as indicator of climatic conditions he demonstrated how cow milk-yield declined as one moved from the drier regions of the north-west to higher rainfall regions in east and south-east India. In the same way, he also noted that the work-capacity of bullocks, as indicated by culturable command area of a pair of bullocks, declined with increasing intensity of rainfall (Burns 1944). Following Burns, Mishra (1978, Chapter 7) formulated and computed a regional livestock production planning model for India, dividing the country into six livestock regions according to rainfall intensity, topography, and territorial contiguity.

Recognizing the determining influence of environment on livestock productivity, however, does not tell us the whole story. It leaves out environmental determination of livestock biodiversity, a feature that has acquired invaluable economic importance in view of the recent revolutionary breakthrough in the technology of livestock breeding and genetic engineering. India's environmental diversity has conditioned its livestock biodiversity, an aspect briefly described in Section 2.2 below.

We noted earlier that the relationship between environment and livestock is interactive. Just as environment has a determining influence on livestock productivity and biodiversity, livestock production system has a feedback influence or effect on the environment. It has been known for ages that overstocking and resultant overgrazing of a pasture beyond its natural carrying capacity is damaging, and eventually destructive of the pasture, not of its plant communities alone but also the faunal life it supports. In India such kind of negative feedback were first articulated and raised to the political level by the environmentalists during the mid-1980s. They alleged that grazing by small ruminants, such as sheep and goat, especially in the fragile zones such the Arravali ranges, was damaging their ecosystems. Prime Minister Rajiv Gandhi quickly got an order issued, stopping financial support and distribution of sheep and goat under the anti-poverty programme. Under the counter-pressure of the state level politicians, the government appointed a Task Force of experts to examine the issue. The group came to the conclusion that the environmentalists' claims were unfounded (GOI 1987). Be that as it may, overstocking and consequent overgrazing does damage pastures and wasteland ecology.

The other negative environmental effect of livestock production systems, which came into prominence during the early 1990s, relates to global warming and climate change. Ruminants, especially the large ones, such as cattle and buffaloes, are a source of methane (CH_4) emission, a greenhouse gas which has many times more weather warming potential than carbon dioxide (CO_2). In the process of decomposition of carbohydrates through fermentation in the animal's rumen, a part of the energy gets converted into methane which the animal releases mostly by eructation. This is what is known as *methane emission due to enteric fermentation*. Excreta of the animals, if it is stored in pits or lagoons and covered up to prevent atmospheric contact, is also a source of methane formation and emission. This is known as *emission due to manure management*.

It was in the context of the Earth Summit at Rio de Janeiro in 1992 that agriculture, especially rice cultivation and livestock, *inter alia*, came to prominence as major sources of methane emission worldwide. Preparatory to the summit, Western governments, especially the US administration, put in a considerable amount of resources for the estimation of greenhouse gases emission country-wise and source-wise around the globe in anticipation that these were going to be the negotiating points for any post-summit protocol on emission reduction. The US Environmental Protection Agency (US EPA) took the lead in this regard. Since then the World Resources

Institute (WRI) based in Washington, D.C. has made it its business to produce an annual directory, giving data on country-wise and source-wise greenhouse gases emission. India, for the simple reason that her livestock holdings are the largest in the world, figures at the top of their list in respect of methane emission from livestock sources. The estimate made for USEPA (Ahuja 1990) and by the WRI are questionable, since these are not based on the required parameters estimated to represent Indian conditions. However, there is no denying the fact that methane emission from livestock sources has a negative global environmental effect. This and other aspects of greenhouse gases emission from livestock are briefly discussed in Section 2.3.

The environmentalists who see livestock as environment damaging and scientists who measure methane emission are both narrowly focused. In contrast to the negative environmental effects of the livestock production systems, there are also a number of positive effects, especially from India's livestock production system, which are neither recognized nor raised at the national or international fora. The proverbial 'cow dung' used either as crop fertilizer or as domestic fuel by millions of rural households in India has notably a positive environmental value. Similarly, India's livestock function as a gigantic recycling machine of agricultural by-products, producing useful products and services, and saving a precious natural resource, land. This sort of positive environmental contributions are briefly described in Section 2.4. A summary of conclusions is given in the last section.

2.2 Livestock Biodiversity Due to Environmental Diversity

India's large territorial size (328.73m ha) and geographical spread from near equator to sub-temperate latitudes, about 8° to 37° N and 67° to 98° E. longitude, her physiography, a variegated mosaic of cold and warm deserts, snow-clad Himalayas, the Deccan plateau bound to the north by the Vindhya and Satpura ranges and skirted around by the Eastern and Western Ghats and the Nilgiris, the mountainous tracts of the north-east region, and the great river valleys draining the land, her varied climatic conditions with extreme differences in temperatures and rainfall in different parts of the country, all these together have made India a country of high environmental diversity. Based on temperature, rainfall, physiography soil-type, and plant growing period, the National Bureau of Soil Survey and Land Use Planning has identified 20 broad agro-ecological regions in the country (National Bureau of Soil Survey and Land Use Planning 1992).

It is in such diverse environmental conditions that the country's livestock have evolved and adapted over thousands of years. And, that diversity has had a role in determining livestock biodiversity. In its simplest form, biodiversity is measured by the number of present species and sub-species of animals and their breed-types. According to the Indian Livestock Census presently there are 10 species of livestock in India: cattle, buffalo, sheep, goat, horses and ponies, donkey, camel, yak, mithun and pigs. Yak and mithun, however, are sub-species of cattle. Diversity is expressed more in the number of breeds the species have. On the basis of current knowledge, cattle, excluding yak and mithun, have as many as 26 well defined indigenous breeds. Buffaloes have no less than 15 breeds, some of them considered as best milk-breeds in the world. Incredible though it may seem, there are as many as 40 breeds of sheep and 20 breeds of goat in the country (GOI 1996).

It is likely that many more breeds, not only of cattle, buffalo, sheep, and goat but also of other species of livestock, will be identified in the future. Our present knowledge is based on the old scientific investigations, which searched for breeds with high economic performance traits. It is not surprising that the identified cattle breeds, for instance, were classified as *milk-breeds*, *draught-breeds,* and *dual-purpose* breeds. A breed may not serve any specified purpose but it will still be a breed provided it has distinct reproductive characteristics not found in others. For example, in the Trans-Himalayan region of Spiti in Himachal Pradesh, a region that was till recently isolated, there is a lightweight, black–red coloured cattle stock which appears to be a distinct breed but is not included in the list of 26 breeds of cattle mentioned earlier. Cows of this stock normally have a lactation period of 2 years. But there are some cows that have lactation periods of up to 3, 4, 5, and 6 years, though they yield petty amounts of daily milk.[1] Their incredibly long lactation length is unique, not found among cattle anywhere else in the country.

[1]Charudutt Mishra, Scientist (Wildlife and Environment), Nature Conservation Foundation, Mysore, India, 2000. Personal Communication.

Biodiversity judged in terms of the number of breeds a species has does not tell the whole story about the genetic variability in the population of the species. The 26 identified breeds of cattle are found in limited numbers in their home tracts across the country. As a guess estimate, it is said that over 80 per cent of the country's cattle population is nondescript, that is cannot be put in pure breed categories. Yet this large nondescript population may show a high degree of genetic variability. A nondescript cow in Haryana, for instance, is a very different animal compared to a nondescript cow in the Assam valley in body structure, weight, and reproductive and productive characteristics. Little is known about the genetic variability of this large nondescript cattle population of the country. So is the case with other species of livestock. It is only recently that the National Bureau of Livestock Genetic Resources at Karnal has initiated a livestock genome project. With revolutionary changes taking place in livestock breeding and genetic engineering, the existing genetic variability, in other words, genetic wealth of the country, has unforeseen possibilities of generating economic wealth, though it is next to impossible to put an economic value on the genetic wealth itself.

The livestock breeding policies pursued so far have not been helpful to the preservation of the biodiversity of the indigenous stock. The policy of crossbreeding nondescript indigenous cows with exotic breeds (Jeresey and Holstein-Freizian) is reported to have endangered such indigenous breeds as Ongole, Haryana, Tharparkar, Gir, and Kankrej due to depressed demand for these breeds (GOI 1996). Kerala's dwarf cow, the *Vechure*, a distinct milk-breed with high fat percentage in milk, had virtually become extinct. It has now been revived (Iype, 1997). Tractorization of agriculture with subsidy support for purchase of tractors may have endangered draught-breeds of cattle in the country. Thankfully, crossbreeding of sheep and goat with exotic breeds did not get off the ground in terms of expected results.

2.3 Greenhouse Gases Emission

The livestock production systems are generally classified into three broad groups: *pastoral*, in which animals are almost exclusively maintained on grazing; *mixed-farming*, in which livestock and agricultural production are integrated; and *industrial*, in which production depends upon external feed supplies, and is market driven. Globally the industrial livestock production system has been gaining ground and pastoral system has been declining. Still in most parts of the world, except in some OECD countries and the US, mixed farming is the norm (de Haan 1997). In India though there are still some pastoral communities like the Gaddis and the Gujjars in the Himalayas, pastoralism has been declining as more and more of land under pasture is taken away for other uses. Industrial livestock production is limited to urban areas. In 1992, only 6 per cent of the total cows and also of buffaloes in-milk were engaged in milk production in urban areas (GOI 1992). The predominant mode is mixed farming supplemented by grazing on accessible neighbourhood pastures, wastelands, and forests. Accordingly, the discussion in the present and the following section on environmental effects relates to the mixed farming system.

There are a number of greenhouse gases emission from the livestock production systems. In addition to methane and carbon dioxide, which are of major importance, there is also emission of trace gases like carbon monoxide (CO) and nitrous dioxide (NO_2). The following subsections provide a brief exposition of the processes and causes of methane and carbon dioxide formation and emission.

2.3.1 METHANE EMISSION: ENTERIC FERMENTATION

As mentioned in Section 2.1 there are two routes to methane formation and emission from ruminant livestock sources: enteric fermentation of feed and manure management. To begin with the first, methane formation occurs in the digestive tract of the animal, the rumen. The animal releases most of the methane thus formed by eructation. Fermentation of the feed in the rumen is a necessary process for decomposition of feed into nutrients and energy. A part of the energy thus generated is used up or 'lost' in the formation of methane through a complex process of chemical reactions.

Chemical analysis of an item in any given feed ration before it is fed to animal, *inter alia*, gives us the quantum of digestible crude protein (DCP) and the carbohydrates from the cellular fibrous (dry and green roughages) and non-fibrous (concentrates) feed components in the ration. Carbohydrates are the source of energy measured in their calorific value. If this given feed ration is actually fed to an animal, the energy equivalent of the carbohydrates contained in the ration is what is called gross energy intake (GEI).

But digestibility of the feed ration is another matter. Not all of the feed fed to an animal is digestible or digested in the sense that all of it is completely converted into utilizable proteins and energy. Part of it passes out as excreta of the animal. Animal nutrition research has established that digestibility of concentrate feeds is higher than of roughage. It is in the process of breaking down the carbohydrates from the two sources of feed during enteric fermentation of feed that a part of the gross GEI gets used up in the formation of methane. Since concentrates have higher digestibility, it means relatively lesser proportion of the gross energy intake is used up in methane formation during enteric fermentation for the same amount of carbohydrates if these were to be derived from fibrous roughages, green or dry. It is now well established that in the US feedlot system of beef production in which feed-grains account for 90 per cent of the ration, the methane share of the GEI on average is almost reduced to half (3–4 per cent) as compared to most other situations in which feed ration consists largely of forages supplemented by concentrates (Johnson and Johnson 1995). In the Indian mixed farming conditions where feed ration largely consists of roughages (green and dry) supplemented by concentrates, methane's share of GEI is expectedly higher. The UN Intergovernmental Panel on Climate Change (IPCC) has reported that in most feeding situations around the world, including South Asia, methane emission as a share of GEI on average is 6 per cent for dairy cattle and 7.5 per cent for all buffaloes, 6 per cent for sheep, and 5.0 per cent for goat (IPCC 1995, Appendix Table A-1, A-3, and A-4). The higher rate for buffaloes is, incidentally, problematic, since buffalo is said to be more efficient in converting roughages than cattle (ICAR 1985). The IPCC figures are based upon research done in the West. The IPCC also gives methane emission rates in kg per animal per year in different regions of the world according to age groups and functional categories, such as dairy-cows and non-dairy cows. The IPCC calls these '*default rates*', recommended for use by countries which do not have or defaulted on having estimates of their own methane emission rates. Though not clearly stated, the default rates for any region, say South Asia, seem to have been computed in the following manner. Assume the average body weight of animals of a species in different age groups. For the average body weight animal in any age group, work out its per day gross energy *requirement* for maintenance and production (growth, milk yield, work for instance). Apply the methane loss rate of the relevant species, noted earlier, to the worked out energy requirement to obtain the quantum of methane loss in joules and then convert it into kg. Multiply this by 365 to obtain methane emission in kg per animal per year.

Notwithstanding the assumption regarding average body weight the above procedure has a major untenable assumption behind it in that it implicitly assumes that gross energy *requirement* of the animal is equal to its gross energy *intake* (GEI). Most studies to date in India have shown that nutritional *requirement* of feed and fodder of the livestock population falls short of the *availability* (Nutrition Advisory Committee, ICAR, 1954 NCA, 1976 Feed and Fodder Committee, Ministry of Agriculture, Majumdar and Singh 1992). And, it may reasonably be argued that feed/fodder *consumption* is less than the availability (see Chapter 3). That being so, IPCC default rates, if applied to India, will overestimate total methane emission due to enteric fermentation from its livestock sources. In view of this we have not used IPCC default rates for estimating methane emission due to enteric fermentation from Indian livestock sources (see Chapter 5).

It may be noted that the discussion above has centred around *averages*, whether it relates to methane emission as a percentage of GEI or IPCC default rates. Experimental research has, however, shown that there is a great measure of variability associated with methane emission *vis-à-vis* GEI. Furthermore, the range of variance of methane emission increases with increase in the digestibility of diet. More interestingly, as the scatter in Fig. 2.1 shows, methane emission ranges from about 2 to over 10 per cent of the GEI in the diet digestibility range of 65 to 90 per cent as indicated by digestible energy (DE) as percentage of GEI. In contrast, the range of methane emission is limited to between about 5.5 to 8 per cent of GEI in the digestibility range of 50 to 60 per cent (Johnson and Johnson 1995).

Note that the high digestibility range corresponds to the dietary conditions in the developed countries and low digestibility corresponds to a country like India where diet consists a great deal more of dry roughages supplemented by greens.

Now, if we take the *simple average* of the end points of the methane emission range, in the former case it works out to 6 per cent [(2 + 10)/2], while in the latter case it works out to 6.75 per cent [(8 + 5.5)/2], that is almost the same. In other words, the *simple*

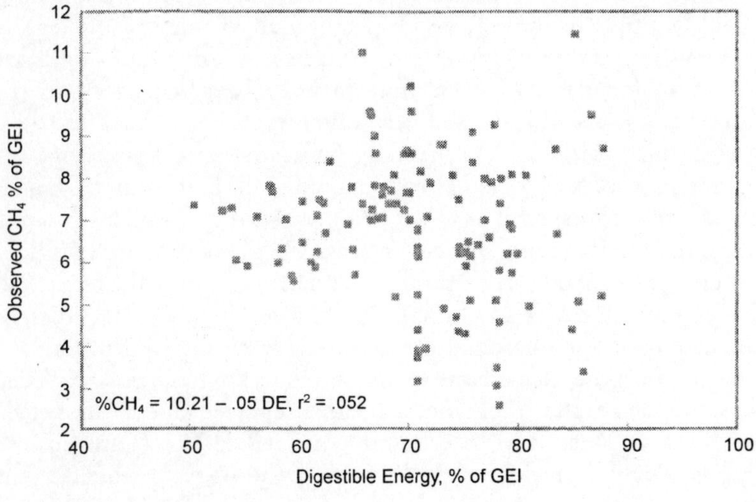

Source: Johnson and Johnson (1995).

FIG. 2.1: METHANE EMISSIONS FROM CATTLE

averages cover up the effects of diet digestibility on the incidence of methane emission. This is what the IPCC has precisely done by putting 6 per cent of GEI as methane emission rate for dairy cattle from enteric fermentation in all the regions of the world (IPCC 1995, Table A-1). In the low digestibility case where the distance between the high and low points of the methane emission range is small (8.0–5.5), the simple average of 6 per cent of methane emission rate may be accepted as close to reality. But in the high digestibility case where that distance is very large (10–2.0), the simple average of 6 per cent methane emission rate is misleading, because the variation here is more than three times compared to the former case. Observed variance apart, measurement of methane from enteric fermentation itself is not without difficulty. There are several measurement techniques applicable to individual animals and also techniques for animals in group such as in pastures. Each has its advantages and disadvantages, of which Johnson and Johnson (1995) provide a good summary. In India, the old gas mask technique has been used to measure methane emission from individual animals so far. Krishna's experimental study (1978) may be cited as an example. We are given to understand that the National Dairy Development

Board (NDDB) is now preparing to take up measurement using non-isotopic tracer technique, which is supposed to be more efficient.

Finally, a number of mitigation strategies for reduction of methane emission from enteric fermentation have been suggested in the literature, especially for the developing countries. These fall into two broad categories: (1) breeding programmes which change the germplasm of the local, indigenous animals in favour of higher production traits, such as those being pursued by India by crossing local cows with exotic breeds, or 'upgrading' nondescript buffaloes by crossing with Murrah buffaloes, and (2) increasing the digestibility of feed. Adding of ionophores like monensin to the diet or defauning of the microbial methenogens from the rumen with the specific purpose of methane emission reduction have not produced sustainable results. Emission reduction reported due to monensin addition is likely due to reduction in feed intake (Johnson and Johnson 1995). In view of this our observations are confined to the two sets of mitigation strategies noted above.

The mitigation strategy of germplasm improvement needs no elaboration. The suggested measures for increasing feed digestibility are the following: (*i*) changing the feed composition to balance the diet of animals by addition of non-fibrous, starchy feeds like beetpulp or feed grains; (*ii*) chaffing, grinding, and pelleting of forages; (*iii*) supplementation with proteins, lipids, minerals, urea, molasses, or Multi Nutrient Blocks, etc.; and finally (*iv*) urea or ammonia treatment of straw.

The argument in favour of both the strategies has been the same. Both increase the output per animal, result in reduced herd size, increased level of output, and reduced methane emission per unit of output. Cees de Haan et al. (1997) have characterized it as a *win-win situation*. However, there are environmental trade-offs of both the strategies. Nature hardly ever offers a win-win result to anthropogenic interventions. As regards the strategy of changing the germplasm, it affects the livestock biodiversity of a country and reduces its genetic variability (see Section 2.2). The Indian experience with crossbreeding of cattle with exotic breeds has shown that crossbreeds have lower conception rates and higher mortality rates. Milk yield is reported to fall in the F2 generation of cows, and the incidence of infertility is reported to increase.

The strategy of increasing feed digestibility by changing the feed composition in favour of non-fibrous carbohydrates such as feed-grains too has an environmental trade-off. Additional feed-grains production, *inter alia*, will require additional land allocation. Land being a natural resource, the less demand that is made on it, the more environment friendly will it be. Therefore, feeding less digestible crop residues such as wheat and rice straw to animals is an environment friendly activity, since it saves on land, and prevents emission of carbon dioxide and trace gases that would otherwise occur if the residues were to be burned (more on this later). Lastly, as we have seen, at high levels of feed digestibility, the methane emission rate has very high variance, and therefore, not necessarily lower emission per unit of GEI.

2.3.2 METHANE EMISSION: 'MANURE MANAGEMENT'

In the IPCC terminology, 'manure management' includes all systems of disposal of animal excreta such as disposal in lagoons and landfills, disposed as dropping by animals in pastures, disposal by burning as dung cake in India, etc. In the Indian practice the term 'manure' refers to that part of the dung output which is actually used as manure, that is as organic fertilizer. The other part of the dung output is used to make dung cake for burning as domestic fuel. Both these systems of disposal of dung get covered under methane emission from manure management.

Methane emission from manure management is much less compared to emission from enteric fermentation. A study carried out in the UK on a 165 head dairy operation found that an average animal generated 105 kg methane annually through ruminant processes and only 0.22 kg methane from dung while dung piles were kept under anaerobic condition (Jarvis et al. 1995).

There are several factors which determine methane emission from manure management. These are: climate; manure management system; dung output (evacuation rate) per animal; volatile solids (VS) contained in it; maximum methane producing potential of the dung output as indicated by methane per unit of volatile solids (Bo); and methane conversion factor (MCF), which is the fraction of methane actually realized out of the potential (IPCC 1995, section on Agriculture, pp. 4.22–4.26). A brief explanation on each of these follows.

The IPCC has divided the world into three climatic zones according to mean temperature: cool (< 15° C), temperate (15–25°

C), and warm (> 25° C). Methane emission is supposed to be highest in warm climate, followed by the temperate and cool climates. Manure management systems make a major difference to methane emission. When dung is stored as liquid in covered lagoons, ponds, tanks, and slurry-pits, as happens to be the practice in the developed countries, a large amount of methane is produced, since it decomposes in anaerobic conditions. The USEPA, quoting Gibbs et al. (1989), has reported that the disposal of manure output (about 15 tonnes a year) of a 600 kg dairy cow in lagoon could result in 120–825 kg of methane, which is about 10 times more methane than that originating in the animal's rumen (EPA 1990; Gibbs et al. 1989). On the other hand, in a study of 165 head dairy operation in the UK, Jarvis et al. (1995) found that while an average animal generated 105 kg of methane annually through enteric fermentation, methane emission from dung per animal was only 0.22 kg, dung having been piled in anaerobic conditions. In general when manure is handled as a solid, for instance stacked or piled up or dropped on pastures and range lands, it decomposes aerobically, and little or no methane is produced.

Dung evacuation rate, that is output per animal per day, depends upon the quantity and quality of feed intake. Since feed intake varies by age groups and functions of the animals, the dung evacuation rate too varies. Dung output is equal to the undigested part of the feed intake. It can also be seen as undigested part of the gross energy intake (equivalent of the feed intake). Thus, dung output is equal to gross energy intake minus digested energy, all measured in quantities of dry matter.

Volatile solids are the degradable organic materials contained by the dung output (dry matter basis). It is with such materials that methanogenic bacteria react to produce methane. Volatile solids is equal to dung output net of its ash content. Maximum methane producing potential of the dung output (Bo) needs little explanation. It is measured in gaseous volume per kg of VS. Chemical analysis of the dung output yield its VS and Bo values. The methane conversion factor (MCF), that portion of Bo actually realized, depends upon the system in which dung output is managed, and the climatic condition. The latest available estimates of MCF according to climate type and manure management system are to be found in IPCC (1995).

If for a given country one has estimates of the above parameters, the types of manure management systems, and the distribution of dung output according to these systems, one can estimate the methane emission from the populations of the animals of any species in different age groups and functional categories. Summing them up will yield total methane emission of a country from manure management. For countries which do not have their own parameter estimates, the IPCC has given what it calls emission factors (EF), that is methane emission per animal per year for dairy and non-dairy cattle, buffaloes, and swine, in different regions of the world and climate zones, including the Indian subcontinent. Applying these emission factors, the 'default rates', so to say, to the respective livestock populations one can get total methane emission from manure management of the country in question.

Our own procedure for estimating methane emission from manure management in India is given in Chapter 5. This is in line with the IPCC recommendation that wherever feasible national assumptions and data should be preferred for the purpose of greenhouse gases estimation, since 'default assumptions and data may not always be appropriate for specific national contexts' (IPCC 1995, Overview 6). As regards mitigation strategy, the usually suggested measure is to capture methane from dung in the form of biogas energy. India has been in the lead for decades in this regard. However, the promotion of installation of biogas plants with subsidy support has, for various reasons, produced limited result.

2.3.3. CARBON DIOXIDE EMISSION FROM BURNING OF DUNG AND EMISSION OF METHANE AS A TRACE GAS

In India, a substantial portion of dung output is converted into dry dung-cake and used as domestic fuel, especially by millions of rural households. Burning of dung-cake results in carbon dioxide emission, beside methane emission, which is covered under methane emission from manure management discussed in the preceding section.

Dung-cake is a biomass fuel, a renewable or sustainable source of fuel unlike non-renewable, unsustainable fossil fuel. Burning of dung-cake in scattered and small scale, as in the Indian situation, is not supposed to make any net addition to the CO_2 inventory of the country. Net CO_2 emission from burning of all types of 'traditional biomass' such as burning of savanna grass, crop residues, firewood, dung cake, etc., is zero, since CO_2 emission from these sources gets reabsorbed by growing plants and trees through photosynthesis during the present and/or the following cycle of biomass growth. However, a certain amount of methane is released as a trace gas on burning a biomass fuel. This needs to be taken into account in the greenhouse gases inventory (IPCC 1995).

2.4 Positive Environmental Contributions of the Livestock Production System

In Section 2.3 our discussion was confined to greenhouse gases emission from the livestock production system, especially emission of methane which makes a negative environmental contribution by way of its effect on global warming and climate change. The present section focuses on the positive environmental contributions of India's livestock production system. There are several of these: (*i*) saving of chemical fertilizers due to dung used as manure and prevention of greenhouse gases emission in the manufacture of the former and their supply at the farm gate; (*ii*) saving of the natural resource, land, due to dung used as domestic fuel in place of firewood; (*iii*) saving of land due to recycling of agricultural by-products as animal feed; and (*iv*) prevention of greenhouse gases emission, especially of CO_2, due to animal draught power used in agriculture. The following subsections provide a brief description of each of these, starting from the last one.

2.4.1 Prevention of Carbon Dioxide Emission due to Animal Energy Use in Agriculture

The interdependence between agriculture and the livestock production system is typically demonstrated by use of animal draught power in agricultural operations—land tillage, lifting water for irrigation, threshing of harvests, haulage of crop residues and grains, transport of men and materials in the rural context. The energy source of the draught power is the feed fed to working animals. Regardless of whether this feed is a by-product or product of agriculture such as cultivated green fodder, it is a renewable and sustainable source of the energy. Working animals are replaceable by tractors and other agricultural machinery run on fossil fuel or electricity generated from fossil fuel. This source of energy, in contrast to animal feed, is non-renewable and unsustainable.

In India, over 70 million working animals are annually used to perform agricultural operations and rural transport. Consider replacing this stock by tractors. The substitute number of tractors will have to be run on diesel. The subsequent burning of the required diesel will cause emission of carbon dioxide and other greenhouse trace gases year after year in future adding to the accumulation of the greenhouse gases inventory. It is the emission of this amount of carbon dioxide and other trace gases and their accumulation that is prevented by the existing stock of working animals.

In order to estimate the CO_2 emission prevented, one needs to know: (*i*) the replacement (substitution) rate between working animals and tractors; (*ii*) diesel required per tractor per year for performing agricultural operations and rural transport; and (*iii*) CO_2 emission per unit diesel. We can then work out the total number of tractors that would be required to replace the existing stock of working animals, total amount of diesel required, and the amount of CO_2 emission that would be caused by burning the diesel in a year.

It is quite common to take an engineering view of substitution between working animals and tractors. A bullock is usually rated equal to 0.5 horsepower (HP). A 25 HP tractor is, therefore, supposed to replace as many as 50 bullocks. Farm level studies on tractorization, however, do not yield such a high rate of substitution. The range of substitution found varies between 4 and 10 bullocks (Mishra and Sharma 1990). Analyses of the change in the stocks of working animals and tractors over time in states like Haryana and Punjab which are advanced in tractorization of agriculture has also yielded substitution rate of similar order (see Chapter 6).

In contrast to the *engineering rate of substitution*, farm level studies, as well as analysis of macro level relative changes in the stocks of tractors and working animals, give us what may be called the *economic rate of substitution*. It is this rate which incorporates or reflects actual farming practices and institutional structure, such as size distribution of landholdings. Economic substitution rate is, therefore, the relevant rate which should be used to find out the number of tractors that would be required to replace the existing stock of working animals, as we have done in Chapter 6.

2.4.2 SAVING OF NATURAL RESOURCE, LAND

As noted earlier, land saving occurs due to dung used as domestic fuel and recycling of agricultural by-products as animal feed. Any activity that substitutes for or replaces another activity that requires land is an environment protecting, nature conserving activity. Both activities, that is using dung as fuel and recycling of agricultural by-products as feed, are of this nature. Brief observations on each of these are in order.

2.4.2.1 *Land Saving Due to Dung Used as Fuel*

Dung used as fuel can be substituted by fossil fuel, like kerosene, or traditional biomass fuel, like fuel wood. The former choice is a priori eliminated, since fossil fuel substitute is non-renewable and unsustainable. The latter choice is reinforced because like dung cake, fuel wood as a energy source is sustainable, and net CO_2 emission from burning of either of them is zero (IPCC 1995).

If dung used as fuel were to be replaced by fuel wood, it will require land for fuel wood plantation from which an equivalent quantity of fuel wood (in terms of thermal energy) could be harvested and supplied to consumers. The area of land required will depend upon two factors: (*i*) the equivalent quantity of fuel wood required each year and (*ii*) the gestation lag between planting and harvesting of fuel wood trees. The longer the gestation period, larger will be land area required to be kept under plantation. The dynamics of this relationship and the procedure for estimating the land area that would be required to be kept under fuel wood plantation year after year is presented in Chapter 4. It is that area which is saved by using the annual dung-cake output of the country as domestic fuel.

2.4.2.2 *Land Saving Due to Recycling of Agricultural By-products as Animal Feed*

If agricultural by-products have no alternative uses, it makes ample economic sense to use them as animal feed since their social opportunity cost is virtually nil (see Chapter 10). But how is it, one may ask, that the same thing is deemed to make a positive contribution to the environment? The answer is simple: it is because recycling of agricultural by-products as animal feed saves on land that would otherwise be required to produce substitute amounts of animal feed. And, saving of any natural resource, in this case land, is an eminently environment protecting activity. It is well known that recycling of waste materials back into the production process is environment protecting to the extent it reduces the demand for extraction of fresh materials from the available stock of inorganic natural resources. In the same way, recycling of agricultural by-products into the livestock production processes is environment protecting to the extent it reduces the need for producing substitute feeds using the natural resource, land.

The feed fed to animals in India is comprised of three parts: (*i*) by-products of agricultural production and processing; (*ii*) cultivated green fodder; and (*iii*) green grasses from grazing. While the first part is the mainstay of feeding, the latter two parts are supplemental. The agricultural by-products are of two types: roughages and concentrates. Roughages consist of crop-residues such as rice straw; wheat *bhusa*; stalks of harvested maize, bajra, jowar, etc. Concentrates consist of grain by-products like husk; bran; oil-cake; grains residues from milling, grinding, or pounding; cotton seed; molasses etc.

If the by-products feeds were to be replaced by cultivated green fodder or feed-grains, either choice will require land for their production, apart from other resources (labour and capital) that will also be required. It is that land area which is saved by recycling the by-products as feed.

We have some basis of comparison between a given quantity of by-products feeds and its potential alternatives. Comparison is possible since all types of feed are convertible into their energy values and/or total digestible nutrients (TDN). The energy value of different kinds of feed can be estimated with the help of energy value documented by Krishna et al. (1978). If the gross energy intake of the bovine population made available from by-products feed, concentrate and dry fodder, were to be replaced by energy from cultivated green fodder, the quantity of green fodder that will be required can be found. Given the yield of green fodder, we can then find the land area that will annually be required to produce the replacement amount of green fodder. The other alternative of producing feed-grains as a potential alternative of by-products feeds is not practised in the country, and its yield parameters, therefore, are not known.

2.4.3 SAVING OF CHEMICAL FERTILIZERS DUE TO DUNG USED AS MANURE

The major portion of dung produced in India is used as manure. It provides plant nutrients: nitrogen, phosphoric acid and potash. Being an organic manure, unlike chemical fertilizers, it protects the soil born micro-organisms and the natural structure of the soil (Garg et al. 1971). If dung manure were to be replaced by chemical fertilizers, the latter will require to be manufactured, making a demand on fossil-fuel energy. Dung manure, on the other hand, is a by-product of the livestock production system making no such demand. The management of dung manure is associated with greenhouse gases emission, CH_4 in particular. The manufacture of equivalent amounts of fertilizers, their distribution, and supply to farm gate will also cause greenhouse gases emission from fossil fuel used. Provided the required data on emission parameters and amount of fossil fuel used were available, the total amount of emission from equivalent amount of fertilizers can be estimated and compared with emission from management of dung manure. This is feasible since different greenhouse gases are convertible into their CO_2 equivalents. We expect that emission from equivalent amounts of fertilizers would be larger. However, for the present our scope of analysis is limited to saving of chemical fertilizers due to dung used as manure.

Dung manure contains all the three nutrients mentioned above. According to a study by the National Council of Applied Economic Research (1959) it contains 0.35 per cent nitrogen, 0.15 per cent phosphoric acid, and 0.20 per cent potash. From a given quantity of dung manure we can find out the respective nutrients. If those nutrients were to be obtained from chemical fertilizers, we ought to first know the nutrient contents of the respective types of fertilizers. Ammonium sulphate, a nitrogenous fertilizer, contains 20.6 per cent nitrogen; superphosphate contains 44.46 per cent phosphoric acid, and mutriate of potash contains 66.0 per cent potash. We can now ask how much of each of these types of chemical fertilizer would be required to yield the same amounts of the respective nutrients contained in the given quantity of dung manure. Suppose x, y, z are the respective amounts of the three nutrients contained in the given quantity of dung manure. Then the respective amounts of chemical fertilizer required to yield x, y, z can be obtained by applying their nutrient content percentages noted above. To illustrate, let X be the quantity of dung manure, which, apart from other nutrients, yields x amount of nitrogen. If this amount of nitrogen were to be replaced by nitrogen from ammonium sulphate, Y quantity of ammonium sulphate will be required such that 35 per cent $Y = x$ where 35 per cent is the nitrogen content of ammonium sulphate. Given x, we can calculate Y, the replacement quantity of ammonium sulphate. We can similarly find out the quantities of the other two fertilizers which will be required to replace the respective nutrients from the given quantity of dung manure, *X*. Given the total quantity of dung used as organic manure in any year in India, we can using the procedure just described to easily calculate the replacement quantities of chemical fertilizers, in other words, the quantities of the fertilizers saved.

2.5 SUMMARY AND CONCLUSIONS

To sum up, India's environmental diversity provides ecological conditions for a high degree of livestock biodiversity. Neither the livestock biodiversity, nor their genetic variability have been fully documented yet. Both are natural endowments, in other words wealth, which ought to be preserved. The livestock breeding policies followed in the country since independence have not been helpful to the preservation of livestock biodiversity and genetic variability, their goal having been circumscribed to productivity gains.

Environmental concerns about the country's livestock, both domestic and foreign, have been confined to the negative environmental effects of the livestock production system(s). Due to its having the world's largest livestock holding, India has been singled out for the largest amount of greenhouse gases emission, in particular of CH_4 from her livestock sources. The environment protecting effects of the country's livestock production system have been generally ignored, though several such effects exist. By using over 70 million working animals in agricultural operations and rural transport, India prevents greenhouse gases emission that would otherwise occur if this stock were to be replaced by agricultural machinery run on fossil fuel. Similarly, dung output used as manure in India saves on chemical fertilizers and prevents thereby greenhouse gases emission from fossil fuels that would occur otherwise in the manufacture, distribution, and supply of the equivalent amounts of fertilizers at the farm gate level. More important is the saving of the natural resource, land, by use of a part of the dung output as domestic fuel, and by recycling of agricultural by-products as animal feed in India. Note that any such activity that substitutes for another activity requiring land is an environment conserving activity. The point is that a comprehensive environmental approach to studying India's livestock production system, which takes into account both the negative as well as positive effects, is required for a better understanding. In the present study an attempt has been made to follow such a comprehensive approach. The following Chapter 3 provides a description of India's livestock production system and the livestock population base on which it rests. The purpose is to lay down the groundwork for subsequent chapters.

3. The Livestock Production System

3.1 LIVESTOCK ECONOMY: AN OUTLINE

India's livestock economy seen in terms of animal numbers is quite large. In 1992, the year for which livestock census data are available, there were 471 million heads of livestock. In this the largest number was that of bovines (289 million), followed by goat (115 million), and sheep (51 million). This vast number is deployed primarily to produce food (milk and meat), fiber (wool), and animal energy for work. However, in the process of production some *joint products* are also produced. These are urine and dung, and in the case of ruminants, methane from fermentation of feed in their rumen. Then, there are a number of *residual products* the system yields annually from natural death of animals and slaughtering for meat. Notable are hides and skins, bone, hair, horn and hoof, tallow, and offals. All these flow out to industrial uses.

By dint of sheer number India produced 58 million tonnes of milk, a little over 3.4 million tonnes of meat (70 per cent from cattle and buffaloes), and 39 million kg of wool in 1992 (GOI 1997). In relation to human population these output numbers are pretty small. The per capita availability of milk, the most important source of animal protein for Indians, was no more than 182 gm (1992–3) per person per day as against the recommended nutritional norm of about 250 gm (see Section 1.1).

Table 3.1 indicates the livestock sector's position *vis-à-vis* the agricultural sector, of which it forms a part, and its share in the national gross domestic product (GDP). Though there are problems about the Central Statistical Organisation's (CSO's) statistics relating to the livestock sector (Mishra 1999), the figures in Table 3.1 may be accepted as orders of magnitude involved.

Table 3.1 shows that the livestock sub-sector's share in the value of output in real terms of the agricultural sector increased from about 19 per cent in 1980–1 to 23 per cent in 1993–4. Its share in agricultural GDP increased from 13 per cent to 20 per cent during

TABLE 3.1: LIVESTOCK SECTOR *VIS-À-VIS* AGRICULTURE AND THE NATIONAL ECONOMY

(*in Rs crores, at 1980–1 prices*)

Particulars	1980–1	1993–4
Value of Output		
Agriculture and allied activities	56,875 (100.00)	86,318 (100.00)
Livestock	10,597 (18.63)	19,627 (22.74)
Gross Domestic Product		
National	1,22,427 (100.00)	2,36,064 (100.00)
Agriculture and allied activities	46,649 (38.10)	70,231 (29.75)
Livestock	5,913 (4.83)	13,912 (5.89)
Livestock share in agriculture and allied activities	(12.67)	(19.81)

Note: Figures in braces are percentages.
Source: CSO, *National Accounts Statistics*, 1993 and 1996.

the same period. Its share in the national GDP marginally increased, while that of the agricultural sector as whole declined from about 38 to 30 per cent. In brief, while livestock's share in national GDP remained stable between 5 and 6 per cent, its share in agricultural GDP substantially increased from 13 to 20 per cent through the period of the 1980s and first half of the 1990s. The change in general has, thus, been in favour of the livestock sector. The reason for this is that with rise in per capita incomes beyond a certain level, food from livestock sources gains in relative importance over food grains in the food preference ordering of the people. That gets translated into a change of the production structure in favour of the livestock sector. This seems to have been happening in India during the 1980s and 1990s.

The livestock economy is important from another angle too, that of the employment opportunity it provides to the people. Unlike agriculture, animal husbandry is a more regular activity through the seasons of the year. For the last three decades it has remained a key element of the governments' anti-poverty programme. Rural poor, landless, and marginal and small farmers are provided credit and subsidy support to acquire productive animals so as to improve their employment and income position. A rough idea of how much employment the sector provides can be had from the following discussion.

The National Sample Survey Organisation (NSSO) periodically conducts surveys on employment and unemployment in the country. From the Report of the NSS 55th Round (July 1999–June 2000) (NSSO 2000), we can get data for number of persons employed in agriculture and allied activities, livestock being one such activity. Suppose that labour absorbed per unit of output in livestock is the same as in agriculture. Then, the proportion of persons employed in livestock will be equal to its share in the value of output (refer to Table 3.1) of the agriculture. According to the above-mentioned NSS Report No. 458 the total workforce consisted of 337 million persons in 1999–2000, out of which 202 million were usually employed in agriculture. Assuming that livestock's share in the value of output of agriculture remained at the level of 1993–4, i.e. 23 per cent, the number of persons employed in livestock activities works out to over 46 million. Note that while the energy output of the working animals is not included in the value of output of the livestock sector, the labour of upkeep of those animals is included in this figure of 46 million. This number indicates the order of magnitude, rather than an exact estimate. Furthermore, it should not be interpreted as if so many persons were engaged wholly in livestock activities round the year. It simply means livestock activities absorbed the labour equal to the work of so many persons for the most part of the year. In addition to the direct employment of this order, millions of people are employed in industries using livestock raw materials, the residual products of the livestock production system, and in the dairy industry.

3.2 Livestock Population Base of the Production System

India accounts for about 16 per cent of the world's cattle population, and more than half of world's buffaloes. As mentioned earlier, in 1992 there were 471 million heads, out of which cattle were 204.58 million and buffaloes 84.21 million in number. Cattle bullocks have been the backbone of Indian agriculture for ages, providing the needed work-energy and sustaining it. They do it even today despite agricultural mechanization having taken place with public policy support. Cows are reared mainly to produce and reproduce the required stock of bullocks. In the process they also produce milk. Buffalo is India's main milk-animal. This sort of functional specialization is reflected in the age–sex composition of the two populations (see Appendix Table 3.1). In 1992, adult females accounted for 31 per cent, and adult males for over 36 per cent of the total cattle stock (indigenous + crossbred). In sharp contrast, adult females accounted for 52 per cent, and adult males only 10 per cent of the buffalo stock. Buffalo's share in the total bovine milk produced in the country ranges between 55 and 60 per cent, despite 40 years of state effort to replace not only indigenous cows but buffaloes too by crossbred cows.

Bovines are not bred for meat purposes in India. As the cow is sacred to Hindus, a devout Hindu cannot bear slaughter of cow and her progeny. Yet slaughter of unproductive, incapacitated, and decrepit cattle does take place. There are no such taboo against buffaloes. There is no reliable official statistics on the number of cattle or buffaloes slaughtered, and the meat outputs from them. It is indeed surprising that the concerned Central government department, even after half a century from independence, relies on the Food and Agriculture Organization's (FAO's) estimates on the rates of slaughtering and, therefore, on meat produced in India not only from cattle and buffaloes, but also from goat and sheep (GOI 1999). But FAO's estimation procedure and estimates of outputs of livestock products in India should be taken with a pinch of salt, indeed with a lump of salt (Mishra and Sharma 1990). Incidentally, the populations of both the species, cattle and buffaloes, have traditionally been adjusted to the requirement by letting the unwanted young borns to die a natural death at the farm level itself.

Goat is the animal reared primarily for meat in India. The stock also produces some milk, conventionally assumed to be between 3 and 4 per cent of the total milk produced in the country. Their number in 1992 (115.28 million) was more than that of buffaloes (84.21 million). Sheep is reared primarily for wool, and usually those in advanced age are slaughtered for meat. Their number in 1992 was about 51.0 million. Goat is a browser, while sheep is a grazer. Both prefer an open environment rather than being kept enclosed and tied to stall feeding. With increasing human demand on pastures and wastelands the scope for having an open environment for goat and sheep has been getting reduced over time.

Contrary to the general impression and environmentalist concerns, India's livestock population has not been growing at an unduly fast rate. The population figures from 1972, 1982, and 1992 are presented in Appendix Table 3.1, which also gives the relative shares of different categories and their annual compound growth rates. Our choice of livestock census years has been guided by two considerations. Firstly, any study of change over a longer period is better, as change over a short period, such as over a 5-year interval between consecutive livestock censuses, can be misleading. Secondly, abnormal years should be kept out of the growth rate exercise. Several of the livestock Census years, 1966, 1977, and 1987, were drought years of severe (1966 and 1987) or moderate (1977) intensity. Livestock, it is well known, become the first casualty of a drought. The years chosen in Appendix Table 3.1 were normal years. Finally, for no good reason, in the 1982 Census the traditional classification of adult cattle males was replaced by new categories, castrated and uncastrated. This vitiates the comparison between censuses as regards working male cattle. Fortunately, the problem is not unsurmountable.

Appendix Table 3.1 shows that the bovine stock, the most important and largest of all categories, grew at a compound rate of 1.02 per cent a year, cattle at 0.68 per cent, and buffaloes at 1.94 per cent during 1972–92. Goat is the only exception among ruminants whose population grew at 2.7 per cent a year during the same period. Others like sheep, horses and ponies, camel, donkey, and yak either remained stagnant or declined in number, posing a threat for survival in the latter case.

The varying growth rates of the different categories of bovines reflect the changes in the composition of the bovine stock arising out of changes in the pattern of demand on the livestock production system. Notable changes are: increase in the proportion of buffaloes, consequent reduction in the proportion of cattle, increase in the proportion of breedable females, and reduction in the proportion of working bovines. The composition has slowly but surely been moving in favour of increasing milk production, and against the production of animal energy. The latter apparently is due to mechanization of agriculture.

The large livestock population of India is dependent on agriculture for feed. Agriculture too is dependent on livestock for animal energy. Section 3.3 presents a brief account of the interdependence between livestock and agriculture. The animal feed availability, nutritional requirement of the population, and consumption of feed is discussed in Section 3.4.

3.3 Livestock–Agriculture Interdependence

The pastoral livestock production system does not depend on agriculture. Neither does the industrial livestock production system in which a production enterprise receives all its inputs externally supplied through the market. It is when an enterprise or a farming unit is organized to produce agricultural crops and livestock products together that we meet with a mixed-farming enterprise. These, at a large scale, with a large number of such enterprises, become a mixed-farming system. This system has been in operation around the world for millennia, since humans made a transition from pastoralism to agriculture. Historically, the three systems of livestock production can also be seen as three phases of general economic development of human society. The industrial livestock production system is a product of industrialization, which since the beginning of the Industrial Revolution in England in the late eighteenth century has depended on energy from non-renewable fossil fuels, coal, electricity and petroleum oil. It is not surprising that the industrial livestock production is in vogue in the developed, industrialized countries of the world. It is only now in the context of the worldwide concern about the environment that the virtues of the mixed-farming system is gaining recognition (de Haan, et al. 1997). Its virtue arises from the fact that it depends upon renewable biomass energy and there is an exchange of energy between agriculture and livestock.

As noted in the preceding chapter, the predominant mode in India presently is that of the mixed-farming system. Tens of millions of rural households practise mixed farming in which agriculture and livestock are interdependent activities. Agriculture provides animal feed, the source of energy and proteins for the livestock. A part of that energy is returned to agriculture by the working animals deployed to perform agricultural operations ranging from ploughing of land to harvesting, threshing, and transporting of the crop produce. Incidentally, a part of the energy received by the working animals is returned outside agriculture in rural transport, in carting of men and materials, especially during agricultural off-seasons. Another way in which part of the energy received is returned to agriculture, not just by the working animals but all others, is in the form of dung used as organic manure for agriculture, dung being undigested part of the energy received. Here again we should take note of the fact that a portion of the dung output in India is converted into dry dung-cakes and used as domestic fuel, especially by millions of rural families. Since our focus of attention is interdependence between livestock and agriculture we have not taken note of the energy and protein which the livestock return to the society in the form of animal food. As a whole, the energy received by livestock from agriculture is in part returned to agriculture, in part used outside agriculture, and in part transformed into livestock food.

Mechanization of agriculture, for whatever reasons, at first weakens the interdependence between agriculture and livestock production, and finally removes it. The two sets of activities become separate, and independent of each other. Livestock production becomes an industry. The production of primary items of animal feed, *feed-grains* in particular, remains an agricultural activity, but the output is for sale rather than for feeding animals at the farm itself. Of course, all this does not happen in one go. There is always a long period of transition from mixed farming to industrial livestock system. The process of delinking agriculture from livestock is simple. Agricultural machinery run on fossil fuel displaces working animals from the farm, the key link between the two. It helps in quicker completion of certain agricultural operations such as ploughing of land and harvesting of crops. Time saving in turn enables farmers to increase cropping intensity through recourse to short duration crops. Displacement of working animals by machines, say tractors, has, however, serious environmental implications. Rarely has any one paid attention to this in the past. We shall return to this problem in Chapter 7.

3.4 Feed Availability, Requirement, and Consumption

3.4.1 Conceptual and Definitional Aspects

Animal feed *availability* and nutritional *requirement* of the livestock population has been a persistent theme of livestock research in India since the early 1950s. Before we take them up for brief comments it is necessary to be clear about the sources of feed and types of feed, and the nutritional functions of feed, and the relationship between feed *production, availability, requirement,* and *consumption.*

Animal feed in India is available from cultivated and uncultivated lands. The uncultivated land consists of the cultivable and un-cultivable wastelands, fallow lands, common pastures and grazing grounds, and accessible (permitted) forest areas. It needs to be emphasized that not all uncultivated lands but only the *accessible areas* where animals can approach to graze and browse or the village inhabitants rearing livestock can approach to gather edible grasses, leaves, barks, etc., are in fact relevant as a source of feed. In short, in this case feed availability depends upon the *accessibility* of the source, apart for the edible flora cover on it during a year (Mishra and Sharma 1990).

The feed available from cultivated lands, in other words, agriculture, is classified into (*i*) products and by-products of agricultural production, and (*ii*) cultivated green fodder. The first group consists of (*a*) grain and grain by-products like oilcakes, bran, husk and items like cotton seeds, molasses, etc., and (*b*) crop residues such as wheat *bhusa*, rice straw, dry stalks of maize, bajra, etc. The availability of both the components depends upon the volume of agricultural output, which in turn depends upon yields of crops and cropped area. The quantity of cultivated green fodder too depends upon the yield and area allocated for fodder crop production. High yielding varieties of fodder crops with irrigation and fertilizers like berseem naturally give higher yields. A notable point is that there is an actual and potential trade-off between availability of feed as a by-product of agricultural crops and feed from fodder crops production. With negligible scope for increasing land area under cultivation as is presently the case in India, a shift of area under fodder crops, *ceteris paribus*, will add to fodder output but there will be a corresponding fall in the by-product feed from other crops. A similar trade-off exists between feed from cultivable wastelands as and when such lands are brought under cultivation.

Two other aspects of feed availability need to be noted. With increasing agro processing industries, manufacture of animal feed has been growing in India. But manufactured feed is not an additional source of feed availability. It is no more than the compounding of available feeds with additives like minerals, vitamins, and chemicals like urea. The other aspect relates to international trade in feed. Tradeable feeds like feed grains and oilcakes and manufactured composites affect domestic availability through export and import. India has for long been an exporter of oilcake and meal, largely to the European Community (EC) countries. To the extent of such exports, net domestic availability is reduced.

Animal feed from all sources is conventionally classified as *concentrates* and *roughages*, and roughages as *dry* and *green*. Dry and green fodder belong to the roughages group, while grain and grain by-products are grouped as concentrates. In order of nutritional value, concentrates are supposed to have higher value than green and dry fodder in this order. Animal nutritionists go a step further, and convert all types of feed into nutrients: protein, energy, fats, minerals, vitamins, and salts. As regards proteins and energy, chemical analysis of a feed gives its crude protein content and its energy value. But not all of its crude protein content or of energy is digestible. Hence, the concept of digestibility comes in. The digestible part of the protein content is termed as Digestible Crude Protein (DCP). The digestible part of the energy is termed Digestible Energy (DE). In the American practice the DCP, and DE from carbohydrates in the feed and fats are summed up to give Total Digestible Nutrients (TDN). In Britain and other European countries, the energy content of a feed is measured as starch equivalent of the feed. From the energy point of view, any feed has carbohydrates of two types: (*i*) nitrogen free soluble carbohydrates (extracts), and (*ii*) carbohydrates from its crude fibre content. But the metabolic process is such that, if required, an animal transforms not only carbohydrates into energy but also proteins and fats. Therefore, the gross energy (GE) content of a feed is derived not only from its carbohydrate contents but also its proteins and fat contents (Sen et al. 1978; Ranjhan 1991).

The nutritional requirement of an animal is made up of two parts: requirement for *maintenance* and *production*. The part that enables the animal to carry on the 'essential life processes while at rest' is referred to as *maintenance requirement*. It depends upon several factors, for instance, body weight, body surface of the animal, and environmental conditions. *Production requirement* is generally considered additive to maintenance requirement. It varies with the level or rate of production (growth of young animal, milk-yield, work intensity). But there is an element of interdependence between the two. A high-yielding milk-breed cow, say a crossbred, requires apart from yield related production ration, a larger maintenance ration compared to a low-yielding indigenous cow of the same body weight. This is because in the high-yielder tissue metabolism is faster and there is more tissue wear and tear (ICAR 1985). Furthermore, if due to better nutrition there is general body-weight gain in the livestock population over time, its maintenance requirement proportionately goes up.

Till the 1970s, nutritional requirements for different categories of Indian livestock were based on scales developed in the West, the most well known being Morisson's scale (Sen et al. 1978). It was in the 1980s that the ICAR Scientific Panel came up with Indian norms and scales by pooling together results of nutritional experiments done in the country during the preceding three decades (Ranjhan 1991). The norms, both in DCP and TDN, are different than those based on the Western scale. Incidentally, the Panel observed that 'buffalo has been found to be more efficient utilizer of crude fiber than the cattle', confirming an age-old folk knowledge.

3.4.1.1 *Relationship Between Requirement, Availability, and Consumption*

The nutritional *requirement* of a livestock population in any year may be more, equal to, or less than the *availability* of feed. On the other hand, actual *consumption* in any year cannot exceed availability net of carry over to the following year. In short, considered *ex post* consumption must equal availability. This condition also provides a check on the validity of consumption and availability estimates. If, *post facto*, the two are not equal to each other then the estimate of either of the two or both are incorrect. Consumption estimate is required not just to check the validity of the availability estimate, but is also important for another reason. All the direct and indirect environmental effects of a livestock production system, as we will see in the following chapters, are related to feed consumption rather than feed availability or requirement.

Most studies in the past have assumed that *consumption* in any year is equal to *availability*, and *availability* is equal to *production* of feed and fodder. There is, however, evidence to show that availability need not be equal to production for two reasons. India has been a substantial exporter of oil cakes to European countries in particular, contradicting the assumption of equality in respect of concentrate feed. And now India will have to open its borders for import of manufactured feed under the World Trade Organisation (WTO) regime. Apart from the effects of international trade on the availability of concentrates, availability is also affected by non-feed uses of agricultural by-products. There are reports that a good portion of soybean oilcake output finds its way to human consumption. Similarly, in some parts of the country rice-straw, otherwise an animal feed, is burnt during the post harvest time and is being increasingly used as a packaging material. The upshot of this is that what is relevant is *net availability*, that is production net of international trade and non-feed domestic uses of agricultural by-products.

3.4.2 ESTIMATION PROCEDURE AND ESTIMATE OF FEED AVAILABILITY AND REQUIREMENT

Given the nutritional norms for maintenance and production for different categories of animal, and the distribution of the livestock population among those categories, it is straightforward to estimate the feed requirement of the population. If the norms are the same, any two estimates may differ only because the assumed levels of production may not the same.

The estimation procedure of feed availability (assumed equal to production of feed) from different sources should again be a straightforward matter. Any two estimates of availability for any given year may, however, differ from each other due to differences in the values of the parameters used for estimation purposes. As noted earlier, there are three sources of animal feed: (*i*) by-products of agriculture; (*ii*) cultivated green fodder; and (*iii*) greens either grazed or gathered from uncultivated lands. Given a set of *conversion ratios* between different grains and grains by-products (concentrates) and between grains and crop residues (dry roughages) one can estimate total availability from this source using total outputs of grains in the year concerned. Similarly, given the yield of fodder crops and the area under fodder crops, the availability of cultivated green fodder can easily be obtained. The most problematic is the estimation of availability from the third source. Nobody really knows the amount of green grazed by animals across India, nor the amount of greens gathered. One way is to assume yield rates for different categories of uncultivated lands and apply it to land use statistics relating to the uncultivated lands to get the availability from this source. But, as mentioned earlier, accessibility of animal to grazing and of village inhabitants to gather feed is a critical factor that should differentiate between what is available (at assumed yield rates) from production and what is actually available from grazing and gathering. In the light of these observations it is not far off the mark to say that an estimate of availability from this source is bound to be woolly.

A number of official committees and individual researchers have made attempts since the early 1950s to estimate feed availability and requirement of livestock population, in most cases the bovine population. The motivation generally was how to increase milk production or meet the minimum nutritional requirement of milk of the human population. Quite often, the estimates of feed availability and requirement are for a year in the future, involving projections of the relevant variables and parameters. In this context, the projection by the National Commission on Agriculture (NCA) of India's livestock population and its composition in precise detail, its feed requirement, and availability from various sources in the year 2000, that is a quarter century away from the base year (GOI 1976), is notable. Incidentally, projections are hardly conformed by the reality. And, the longer the period of projection, more distant from reality it turns out to be, and, therefore, more futile.

Be that as it may, no two studies are comparable since they relate to different years, in addition to the differences in their estimation procedure and the values of relevant variables and parameters used in preparing the estimates. The major studies up to the end of the 1980s are briefly reviewed in Mishra and Sharma (1990), and the notable studies in the 1990s are those by Pandey (1995) and the World Bank (1999). Since the World Bank study is the first such study by an international agency, it specifically deserves some comments. Contrary to the usual run of Indian studies it has prepared what it calls India's 'feed balance estimates' for 1995, in which the term feed 'demand' and 'supply' are used instead of feed requirement and availability, giving the impression that the estimates have some thing to do with market processes. A closer examination, however, suggests that their 'supply' is nothing but availability assumed to be equal to production. And, the 'demand' is some sort of requirement of the projected livestock population (large and small ruminants and pigs) including poultry at the assumed rates of requirement of concentrates and roughages per animal per day. The basis of these rates have not been explained. However, an interesting result of the study is that in 1995, India was *surplus in concentrates* and had a *large deficit in roughages* (World Bank 1999, Statistical Annex Table 3.2C). Almost all Indian studies, on the contrary, have concluded that there was deficit in both, and more so in concentrates.

There are two Indian studies, which stand out of the long line of studies since the early 1950s, which need some discussion since they are directly relevant to the present work. These are by Amble et al. (1965) and by Jain and Singh (1990). In both these studies, *consumption* of different types of feed per day per animal of the different categories of bovines, rather than the *nutritional norms* were, *inter alia*, used for estimating *requirement* of the bovine stock (projected) in the concerned year. The primary source for arriving at the all-India feed consumption rates in both these studies are sample surveys, including repeat surveys conducted by the Indian Agricultural Statistics Research Institute (IASRI) since the second half of the 1950s to the early 1980s, mainly to find out milk-yields and feed consumption per animal per day for different categories of bovines. The surveys were conducted in the typical livestock tracts of the country. Using the results of the 1950s surveys, Amble made the pioneering effort of preparing all-India feed consumption rates for different categories of bovines. He applied the results from the surveyed area in a state similar to the neighbouring but unsurveyed state, and then suitably aggregated the two sets over the major states to obtain average at all-India rates. It seems that Jain and Singh have followed a similar procedure. But in arriving at the all-India rates they have used a long series of the IASRI surveys/repeat surveys conducted over a period of 20 years from 1963 to 1983. Both sets of all-India consumption rates (See Appendix Tables 3.2 and 3.3) involve wide assumptions. In particular, those by Jain and Singh cannot be said to relate to any particular year in the long span of 20 years. However, in comparison to Amble's rates, in every item of feed they are higher. This is as expected for two reasons. Most of the surveys that Jain and Singh draw upon belong to the decade of the 1970s, the high period of the Green Revolution. Secondly, agricultural production increased at a tend rate of 3.0 per cent a year over the period 1966–85, and accordingly, by-product feed output also increased at the same rate. Since bovine stock was growing at about 1.0 per cent a year, per animal feed availability also increased (Mishra and Sharma 1990). In the light of these observations we propose to adopt Jain and Singh's all-India feed consumption rates and apply them to the bovine population of 1992 for purposes of estimating environmental effects of India's livestock production system. The *uncertainty* surrounding the consumption rates can be handled by changing the rates upward and downward and estimating the *range* of a particular effect, rather than wholly relying on a *point* estimate.

3.5 Feed Input–Output Flow

The various dimensions of animal feed have been discussed above in some detail since feed is the primary input the livestock production system receives from the environment, and all its direct and indirect environmental effects are ultimately dependent on the feed consumed by the animals. Figure 3.1 provides a visual idea of the flows involved between the environment and the system. The system receives feed input from three sources. On the right hand side there too many arrows in different directions indicating diverse output flows from the system. Energy flow-back from the system to agriculture via work-energy and dung used as manure has been indicated. Figure 3.1, however, does not display the energy flow-back to pastures by way of animal droppings (dung and urine). Also not shown are the indirect environment protective flows from the system, namely annual land saving due to by-products of agriculture used as animal feed and dung used as domestic fuel; saving of chemical fertilizers due to dung used as organic manure; and, CO_2 emission annually prevented by animal energy used in agriculture. The following Chapter 4 presents a formal model taking into account both the direct and indirect environmental effects of India's livestock production system.

3.6 Summary and Conclusions

India's livestock sector has an important place in the national economy. Its share in the national GDP has remained constant over the years at between 5 and 6 per cent while that of agriculture as whole, of which it forms a part, has been declining. Its share within agricultural output and GDP has been increasing, since the relative importance of animal foods *vis-à-vis* food grains has been increasing with growing per capita incomes in the country. Its share in the 'agriculture allied activities' output valued at 1980–1 prices increased from 19.0 per cent in 1980–1 to 23.0 per cent in 1993–4. Correspondingly, its share in the 'agricultural and allied activities' GDP increased from about 13.0 per cent to 19.0 per cent between

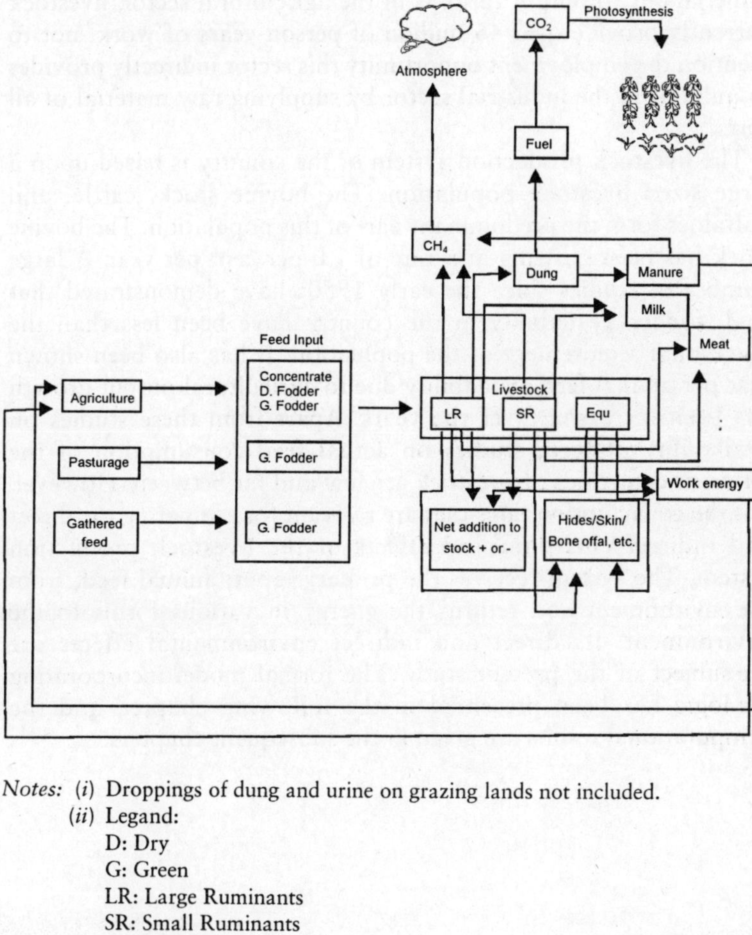

Notes: (*i*) Droppings of dung and urine on grazing lands not included.

(*ii*) Legand:

D: Dry

G: Green

LR: Large Ruminants

SR: Small Ruminants

Equ: Equines

CH_4: Methane

CO_2: Carbon dioxide

FIG. 3.1: FEED INPUT–OUTPUT FLOW CHART

1980–1 and 1993–4. The livestock sector provides employment and income earning opportunities to millions of people. Assuming the same labour to output ratio as in the agricultural sector, livestock currently provide over 46 million of person-years of work, not to mention the employment opportunity this sector indirectly provides to millions in the industrial sector by supplying raw material of all sorts.

The livestock production system of the country is based upon a large sized livestock population. The bovine stock, cattle, and buffaloes form the predominant part of this population. The bovine stock has been growing at a rate of 1.0 per cent per year. A large number of studies since the early 1950s have demonstrated that feed, fodder availability in the country have been less than the nutritional requirement of the population. It has also been shown that per animal feed availability due to agricultural output growth has been increasing over the years. Apart from these studies on availability of feed, studies on actual feed consumption of the different categories of livestock are few and far between. However, it is the consumption rates that are relevant for estimating the direct and indirect environmental effects of the livestock production system. The system receives the primary input, animal feed, from the environment and returns the energy in various forms to the environment. Its direct and indirect environmental effects are the subject of the present study. The formal model incorporating the logic has been presented in the following chapter, and the computational results are given in the subsequent chapters.

APPENDIX TABLE 3.1: NUMBER OF CATTLE, BUFFALO, AND TOTAL BOVINES IN INDIA, 1972, 1982, AND 1992

(*thousand numbers*)

	Year					
	1972	% to total	1982	% to total	1992	% to total
	(a)	(b)	(c)	(d)	(e)	(f)
I. Cattle cross-breds						
A Male over 2.5 years:	-	-	2905	1.52	2,622	1.38
(*i*) Used for breeding only	-	-	-		175	0.09
(*ii*) Used for work only	-	-	-		2,100	1.12
(*iii*) Used both for working and breeding	-	-	-		256	0.13
(*iv*) Others	-	-	-		91	0.04
B Young stock male	-	-	1276	0.67	2036	1.07
(*i*) Under 1 year	-	-	701	0.37	1157	0.61
(*ii*) 1 to 2.5 years	-	-	575	0.3	879	0.46
Total male	-	-	4181	2.19	4658	2.45
C. Female over 2.5 years	-	-	3098	1.62	6493	3.43
(*i*) In-milk	-	-	1835	0.96	4011	2.12
(*ii*) Dry	-	-	949	0.49	1781	0.94
(*iii*) Not yet calved	-	-	299	0.16	575	0.3
(*iv*) Others	-	-	15	-	126	0.07
D. Young stock female	-	-	1884	0.98	4064	2.15
(*i*) Under 1 year	-	-	1012	0.53	2158	1.14
(*ii*) 1 to 2.5 year	-	-	872	0.45	1906	1.01
Total female			4982	2.6	10,557	5.58
Total cross-bred (A+B+C+D)			9163	3.51	15,215	7.44
II. Indigenous						
A. Male over 3 years	74,807	41.94	71,637	37.38	71,867	37.95
(*i*) Used for breeding only	388	0.22	71,375	37.25	10,042	5.3
(*ii*) Used for work only	70,904	79.75	-	-	52,885	27.93
(*iii*) Used both for working and breeding	1997	1.12	-	-	7,990	4.22
(*iv*) Others	1518	1.02	262	0.13	950	0.5
B. Young stock male	22,683	12.72	24,382	12.72	25,071	13.24
(*i*) Under 1 year	11,102	6.22	10,723	5.6	10,832	5.72
(*ii*) 1 to 3 years	11,581	6.49	13,659	7.12	14,239	7.52
Total male	97,490	54.66	96,019	50.1	96,938	51.19
C. Female over 3 years	56,618	31.74	56,804	29.64	57,868	30.56
(*i*) In-milk	22,071	12.37	25,094	13.09	27,551	14.55
(*ii*) Dry	26,442	14.82	26,514	13.84	24,450	12.91
(*iii*) Not yet calved	5078	2.85	4651	2.43	4321	2.28
(*iv*) Others	3027	1.7	545	0.28	1555	0.82
D. Young-stock female	24252	13.6	29,641	15.47	34,563	18.25
(*i*) Under 1 year	11,226	6.29	13,247	6.92	14,924	7.88

(contd.)

(*ii*) 1 to 3 years	13,026	7.3	16,394	8.55	19,639	10.37
Total Female	80,870	45.34	86,445	45.11	92,431	48.81
Total Indigenous (A+B+C+D)	178,360	100	182,464	69.97	189,369	92.56
Total cattle	178,360	75.66	191,627	73.48	204,584	70.84
CGR (%) (1972–92)						0.68
III Buffaloes						
A Male over 3 years	8123	14.16	8278	11.97	8086	9.6
(*i*) Used for breeding only	223	0.39	659	0.95	516	0.61
(*ii*) Used for work only	7059	12.31	-	-	5011	5.95
(*iii*) Used both for working and breeding	607	1.06	6212	8.99	2346	2.79
(*iv*) Others	234	0.41	1407	2.03	213	0.25
B Young stock male	6992	12.19	7999	11.57	9265	11
(*i*) Under 1 year	4429	7.72	4628	6.69	5535	6.57
(*ii*) 1 to 3 years	2563	4.47	3371	4.88	3730	4.43
Total male	15,115	26.35	16,277	23.54	17351	20.61
C Female over 3 years	29,297	51.08	34,679	50.16	43,806	52.02
(*i*) In-milk	15,081	26.29	19,274	27.88	25,892	30.75
(*ii*) Dry	11,338	17.77	12,799	18.51	14,383	17.08
(*iii*) Not yet calved	2250	3.9	2422	3.5	2902	3.45
(*iv*) Others	628	1.09	184	0.27	629	0.75
D Young stock female	12,940	22.56	18,177	26.3	23,048	27.37
(*i*) Under 1 year	6506	11.34	8785	12.71	12,014	14.27
(*ii*) 1 to 3 years	6434	12.22	9392	13.59	11,034	13.01
Total female	42,237	73.65	52,856	76.46	66,854	79.39
Total Buffaloes	57,352	24.33	69,133	26.52	84,205	29:16
CGR (%) (1972-92)						1.94
Total Bovine	235,712	100	260,760	100	288,789	100
(I+II+III)CGR (%) (1972–92)						1.02
IV Miscellaneous						
(*i*) Sheep	39,993	34.02	48,765	30.99	50,783	27.89
(*ii*) Goats	67,518	57.44	95,255	60.54	11,5279	63.32
(*iii*) Horses and Ponies	942	0.8	900	0.57	817	0.45
(*iv*) Mules	75	0.06	131	0.08	193	0.12
(*v*) Donkeys	987	0.83	1024	0.65	967	0.53
(*vi*) Camels	1109	0.94	1078	0.68	1031	0.57
(*vii*) Pigs	6896	5.87	10,071	6.4	12,788	7.02
(*viii*) Yaks	44	0.04	128	0.08	58	0.03
(*ix*) Mithuns	-	-	-	-	154	0.08
Total	117,555	33.32	157,352	27.63	182,070	38.76
Total Livestock	352,767	100	418,112	100	470,859	100

Notes: No enumeration of cross-breed of male and female cattles in 1972; in 1992 one more classification shown, namely castrated and non-castrated for breeding only; in 1982, Male over 3 years belonging to castrated and non-castrated. No classification of A, B, C categories.

CGR: Compound growth rate.

Sources: Livestock Census, 1972, 1982 and 1992; Mishra and Sharma 1990.

APPENDIX TABLE 3.2: OVERALL AVERAGE FEED AVAILABILITY PER ANIMAL/DAY, ALL-INDIA

Classification of Animals	Category	Concentrates (gm)								Roughages (kg)		Nutrient (kg)	
		Oilcakes	Bran	Husk	Cereals	Pulses	Cotton seed	Others	Total (kg)	Green	Dry	DCP	TDN
In-Milk	Cow	34	34	32	13	58	107	10	0.288	4.444	3.54	0.143	2.175
	Buffalo	110	36	128	47	141	307	29	0.797	6.784	5.901	0.261	3.776
Dry	Cow	7	7	105	-	5	-	-	0.125	2.754	2.825	0.071	1.56
	Buffalo	15	5	35	1	37	9	1	0.103	4.316	4.011	0.117	2.226
Adult males	Cattle	35	27	138	8	74	17	9	0.307	4.961	5.652	0.148	3.109
	Buffalo	36	19	65	-	11	12	-	0.174	6.51	5.438	0.178	3.112
Young stock	Cattle	7	5	30	-	3	1	1	0.046	1.581	1.474	0.042	0.821
	Buffalo	3	1	4	-	5	1	1	0.014	1.592	1.657	0.041	0.88

Source: Amble (1965).

APPENDIX TABLE 3.3: AVERAGE FEED CONSUMPTION PER DAY PER ANIMAL

Classification of animals	Category	Concentrates (gm)								Roughages (kg)		Nutrient (kg)	
		Oilcakes	Bran	Husk	Cereals	Pulses	Cotton seed	Others	Total[b] (kg)	Green[a]	Dry	DCP	TDN
In-milk	Cow	96	22	16	14	38	19	68	0.273	5.09	5.01	0.235	3.49
	Buffalo	398	81	28	39	89	181	131	0.947	8.77	6.41	0.373	5.24
Dry	Cow	40	12	3	1	7	1	17	0.081	3.96	4.38	0.131	2.69
	Buffalo	89	58	9	9	28	5	35	0.233	6.84	5.22	0.158	4.49
Adult males	Cattle	58	11	5	55	125	6	17	0.277	5.9	5.48	0.251	3.56
	Buffalo	36	19	65	-	41	12	-	0.174	6.51	5.44	0.178	3.11
Young stock cattle	Male	17	3	2	9	23	1	5	0.06	1.95	1.97	0.089	2
	Female	13	2	1	1	6	-	6	0.028	1.61	2.36	0.059	1.98
Buffalo	Male	1	-	-	-	12	7	3	0.023	1.8	1.53	0.056	1.52
	Female	1	-	-	5	38	5	-	0.049	2.19	2.55	0.082	2.13

Notes: (*a*) Excludes the quantity consumed through grazing.
(*b*) Includes salt, gur, methi, oil, cattle feed, etc.

Source: Jain and Singh (1990).

4. An Environmental Model of the Livestock Production System

4.1 Preliminary Observations

The environment–livestock relationships, the characteristics of India's livestock production system, and its direct and indirect environmental effects have already been described in the preceding two chapters. In the present chapter we are concerned with formulation of a formal framework, a model for determining those effects. The model relates to large ruminants, cattle and buffaloes. The two together, account for over 90 per cent of India's livestock population in terms of equivalent cow-units (Gadgil 1947; Mishra and Sharma 1990). Although the model has been formulated for the cattle stock only, *mutatis mutandis* it applies to the buffalo stock as well. The cattle population's classification as given in the Indian Livestock Census has been reclassified here for our purposes. The first broad reclassification is: (*i*) young cattle, males and females (< 3 years of age); (*ii*) adult females ($\geq$ 3 years of age); and (*iii*) adult males ($\geq$ 3 years of age). The adult females are further reclassified into two groups: (*i*) cows in-milk and (*ii*) dry cows. The dry cows include three census categories: (*a*) dry cows as such; (*b*) adult females not calved even once; and (*c*) 'others'. The adult males are reclassified to give us the population of working males, which includes two census categories: (*a*) males used for work only and (*b*) males used for work as well as breeding.

A noteworthy point is that the dry cows category includes a census category called 'others'. Similarly, the adult males includes a census category called 'neither used for work nor breeding'. Both categories consist of economically useless cattle, comprised decrepit, old animals maintained on scraps of leftover feed on religious or mercy considerations, waiting to meet their natural death on the date of census enumeration. Including them means treating them at par with other animals in terms of feed intake, though in point of fact their feed intake is negligible. Methane emission from enteric fermentation accountable to such animals will have to be netted out from the totals of the categories in which they are included.

The model is a static one. All variables and parameters involved relate to a point of time, a particular year. The model can be made dynamic by introducing a cattle (buffalo) population regeneration sub-model. But this is beyond the scope of the present work. With these preliminary observations, let us now turn to the formulation of the model.

4.2 Methane Emission

4.2.1 Emission from Enteric Fermentation

Let us consider methane emission from the young cattle stock. All animals in the group are supposed to be alike, and they are fed the same quantity and quality of feed. Feed consists of three types: (*i*) concentrates; (*ii*) dry fodder; and (*iii*) green fodder. In order to derive the energy content of the feed, these three types of feed have to be reduced to their respective dry matter equivalents. It is said that energy value of all types of feed on dry matter basis, irrespective of their quality, is the same (4.4 M cal/kg).[1] Yet differences have been reported in the calorific values of concentrates, dry and green fodder (Krishna et al. 1978).

Given the quantities of different types of feed consumed per young cattle per day, their respective dry matter equivalents, and energy values per unit, we can easily calculate the gross energy intake per animal per day. Methane loss in enteric fermentation of feed is found to be, on average, 6 per cent of the gross energy intake in most cattle feeding situations (see Chapter 2). Experiments conducted in India are reported to yield the same average loss rate of 6 per cent in the case of buffaloes too, rather than 7.5 per cent as suggested by the IPCC (1995).

To return to methane emission from young cattle stock, let y denote a young cattle; f_{jy} (j = 1, 2, 3) denote, respectively, concentrates, dry fodder, and green fodder fed to the animal per day; d_j (j = 1, 2, 3) their dry matter fractions; and e_j (j = 1, 2, 3) the energy value per unit of the jth feed on dry matter basis in M Joule (M =

[1]Personal communication with S.K. Ranjhan of the Indian Council of Agricultural Research, New Delhi.

1 million). Then the gross energy intake of the animal per day is obtained as

$$gei_y = \sum_{j=1}^{3} f_{jy} d_j e_j \text{ , } (j = 1, 2, 3) \tag{4.1}$$

where gei_y denotes gross energy intake of the young cattle. The methane emission is assumed to be 6 per cent of the gei_y. Assuming that the animal receives the same amount of gross energy each day of the year, the emission per animal per year is given by the following equation

$$CH_{4y} = 0.06\ [gei_y]\ 365 \tag{4.2}$$

Note that CH_{4y} is in million (M) Joule. M Joules of methane converted into kg gives the emission in weight, the conversion rate being 55.65 (M Joules) = 1.0 kg. Applying this to eqn (4.2) we get methane emission in kg per animal per year as,

$$CH_{4yw} = \frac{1}{55.65} \times 0.06\ [gei_y]\ 365 \tag{4.3}$$

where CH_{4yw} stands for methane emission in weight per young cattle per year, which is referred to by the IPCC as 'methane conversion factor'. Multiplying the total population of the young cattle stock with the conversion factor as given by equation (4.3) gives us the total CH_4 emission from the stock. If Y is the population of the young cattle, total methane emission from Y during the year due to enteric fermentation is

$$TE_y = CH_{4yw} \times Y \tag{4.4}$$

For determining the total methane emission from adult cattle males and females we can write a set of four analogous equations for each of the sub-categories of the males and females. For instance, if the adult female stock is classified into in-milk and dry cows, we can write a set of four analogous equations (4.1–4.4 above) each for cows in-milk and dry cows. Note, however, that apart from notational changes the only parameter that will change from one set to the other is the quantities of intakes of various types of feed, f_j (j = 1, 2, 3), and consequently the gross energy intake (*gei*) per day per animal. The other parameters, d_j and e_j, are constants. Therefore, in order to avoid burdening the text with too many analogous equations, we propose a short-cut for expositional purposes without undermining the necessity of writing all the analogous equations for computational purposes.

For any sub-category of the male and female stock, we determine the 'methane conversion factor' beforehand through analogous equations (4.1) to (4.3). Then, a simple multiplication of the population in any sub-category of males or females with its respective methane conversion factor will yield total methane emission due to it.

Let c_i denote an animal in the ith category of adult females ($i = 1, 2, \ldots n$); CH_{4c_iw} be the methane conversion factor of the ith category, and F_i its population, then methane emission from the female stock as a whole is given by the equation,

$$TE_c = \sum_{i=1}^{n} CH_{4c_iw} F_i \quad (i = 1, 2, \ldots, n) \tag{4.5}$$

where the subscript c on TE denotes adult cows. We can similarly write the equation for the male cattle stock as

$$TE_m = \sum_{i=1}^{n'} CH_{4m_iw} M_i \quad (i = 1, 2, \ldots, n') \tag{4.6}$$

where the subscript m_i denotes adult male in ith sub-category and m total adult male, M_i the population of the ith sub-category, and CH_{4m_iw} its methane conversion factor. Summing up the emissions from the young stock, adult females, and adult males we get the total emission from the cattle population. Let CTE denote total emission, then

$$CTE = TE_y + TE_c + TE_m \tag{4.7}$$

where capital C prefix on TE denotes cattle stock. Given the values of the parameters such as f_i for different categories of the cattle stock, d_j and e_j, and the variables Y, F_i and M_i, the CH_4 emission from the cattle population due to enteric fermentation can be easily calculated.

4.2.2 EMISSION FROM 'MANURE MANAGEMENT'

The second source of methane emission from the ruminant livestock is their excreta, or their dung output. The way the dung output is managed, in other words, stored or disposed, makes a substantial difference to the volume of methane emission. The difference essentially arises from whether the dung output is stored or disposed in aerobic conditions, that is in the open, in contact with atmospheric oxygen, or in anaerobic conditions. In the first case, methenogenic bacteria which feed on or react with the volatile solids (VS) in the dung and produce methane are given little or no chance to produce it. Note that VS are the degradable organic matters in the dung. In the second case they have a good chance to produce methane. According to IPCC (1995), aerobic 'manure management systems' like dung droppings in pastures, spreading dung on field, or storing it in solid form on the open ground or in shallow pits produces 'little or no methane'. In the anaerobic management systems, prevalent in the developed countries, such as storage in lagoons or underground liquid slurry tanks, methane emission is incomparably higher. For instance, whereas in the former case, methane emission ranges from 1 to 2 per cent of the VS in the dung, its range is from 65 to 90 per cent in the latter case (IPCC 1995). The IPCC denotes this as 'methane conversion factor' (MCF) and provides estimates of MCF for different 'manure management systems'. *Ceteris paribus*, climatic conditions, that is whether the climate is cool (< 15 °C), temperate (15–25 °C), or warm (>25 °C), also makes a difference: a warm climate generally has a higher incidence of methane emission for the same management system in comparison to cool or temperate climates.

In India, dung output is almost wholly managed in aerobic conditions, be it the dung stored as solid in the open by the farmer for subsequent use as organic fertilizer, or dung converted into dung-cake for later use as fuel, or dung dropped by animals during grazing on pastures, wastelands, and in forests. For this reason, methane emission from manure management in India should be expected to be small. It is still, however, additive to the total emission from livestock sources. Accordingly, we provide below the formal framework for its determination. Before that it may be noted that when dry dung-cake is burned for fuel, as is the practice in India, along with CO_2 emission there is some CH_4 emission in the process of combustion. This is a separate issue that has been dealt with in Section 4.4.

As in the case of CH_4 emission due to enteric fermentation, consider again an animal belonging to young cattle stock. Its per day dung output will depend upon the quantity of feed intake and its digestibility. The lower the digestibility of the feed intake the larger will be the undigested part, in other words, the dung output per day, which we in India call 'dung evacuation rate'. As noted earlier, methenogenic bacteria operate on the VS in the dung output for producing methane. The VS is nothing other than the animal's dung output on dry matter basis net of its ash content. Ash contents of various types of Indian feeds are available from their chemical analysis (Krishna et al. 1978; Sen et al. 1978; Ranjhan 1991). Let d^*_y be the per day dung output of the animal, α the dry fraction, and β the fraction of ash content. Excretion per day, VS_y is then given as

$$VS_y = d^*_y \times \alpha \times (1\text{-}\beta) \qquad (4.8)$$

where subscript y denotes young cattle. In the absence of data on dung evacuation rate and the dry matter fraction, VS_y can be determined in an alternative way proceeding from the feed intake on dry matter basis. Refer back to equation (4.1). Let dry matter feed intake of the animal per day be F_y, then

$$F_y = \sum_{j=1}^{3} f_j d_j \; (j = 1, 2, 3)$$

where subscript j denotes different types of feed. If r is the feed digestibility ratio, then the undigested portion of the feed, evacuated as dung on dry matter basis, is

$$F'_y = F_y (1\text{-}r)$$

By multiplying this equation by (1-β) for netting out the ash content we get

$$VS_y = (1\text{-}\beta)\, F'_y = F_y (1\text{-}r)(1\text{-}\beta) \qquad (4.8')$$

The estimated value of VS_y from equations (4.8) and (4.8′) should be the same provided dung output and feed intake per day per animal relate to the same point of time and the digestibility parameter r has been correctly estimated. The VS in the dung output has a 'maximum methane producing potential', not all of which is actually realized. Realization depends upon the 'manure management system(s)'. As we have noted earlier in an aerobic management system like the one in India the realization percentage, in other words, 'methane conversion factor', is low. The maximum methane producing potential of a unit of VS is measured in volume of the gas, i.e. cubic meter per kg of VS. With this background, let θ, be the cubic meter value of CH_4 per kg of VS, denote the maximum potential, and let ϕ be the realization fraction, i.e. 'methane conversion factor'. Then per day per young cattle realized volume of CH_4 in cubic meter is ($VS_y \times \theta \times \phi$) where VS_y is obtained from equation (4.8). This, when multiplied by 365 and converted into kg, gives us the per young cattle per year CH_4 emission from its 'manure management'. In other words,

$$CH_{4ysw} = 365\ (VS_y \times \theta \times \phi) \times 0.67 \qquad (4.9)$$

where the multiplier 0.67 is the conversion factor from volume to weight (0.67 kg = 1.0 cubic meter of CH_4), and subscript s denotes emission from a manure management system. The total emission per year from the young cattle stock is then obtained as

$$TE_{ys} = CH_{4ysw} \times Y \qquad (4.10)$$

We can write equations analogous to equations (4.8) to (4.10) for all the different categories of the adult cattle females and males. As we did in the preceding section let us skip writing those equations and suppose that from equations analogous to equation (4.9) we get emission per year per adult female of different categories as $CH_{4c_i sw}$ (i = 1, 2, ... n). The total emission per year from the dung output of the cattle female stock is then

$$TE_{cs} = \sum_{i=1}^{n} CH_{4c_i sw} \times F_i \ \ (i = 1, 2, ..., n) \qquad (4.11)$$

where i = (1, 2, ... n) denotes different categories of adult females and F_i their respective populations. Similarly, the total emission from different categories of the adult males is

$$TE_{ms} = \sum_{i=1}^{n'} CH_{4m_i sw} \times M_i, \ (i = 1, 2, ..., n') \qquad (4.12)$$

The total emission in terms of weight from the dung output of the cattle stock as a whole is given as

$$CTE_s = TE_{ys} + TE_{cs} + TE_{ms} \qquad (4.13)$$

The total CH_4 emission in terms of weight, both from enteric fermentation and manure management, for the cattle stock is obtained by adding equations (4.7) and (4.13), so that

$$CTE_{CH_4} = CTE + CTE_s \qquad (4.14)$$

where CTE_{CH_4} stands for total CH_4 emission from the cattle stock. Given the values of the different parameters involved, the cattle population, and its distribution in various categories we can estimate the total CH_4 emission from enteric fermentation from young cattle stock through equations (4.1)–(4.7) and their analogues in respect of different categories of females and males; and from manure management from young cattle through equations (4.8) to (4.10) and their analogues in respect of the different categories of males and females. The rest is a matter of addition.

4.3 Carbon Dioxide Emission Prevention Due to Dung Used as Manure

As noted in Chapter 2, part of the dung output in India is used as manure. To that extent it substitutes for chemical fertilizers. If dung manure were to be replaced by chemical fertilizers, there would be greenhouse gases emission, of CO_2 in particular, associated with the combustion of fossil fuel used in the manufacture, distribution, and supply of the equivalent amount of fertilizers at the farm gate, which is prevented by dung used as manure. In addition to providing plant nutrients (N, P, K), dung, being organic manure, is not at all harmful to soil-born micro-organisms unlike chemical fertilizers, and also protects the natural fertility of the soil. Ignoring this aspect, let us consider the equivalence between the two in terms of plant nutrients.

According to an old manual (ICAR 1959), *wet dung* contains 0.35 per cent nitrogen (N), 0.15 per cent phosphoric acid (P_2O_5), and 0.20 per cent potassium oxide (K_2O). Goel et al. (1973) used these proportions to estimate amounts of chemical fertilizers that would yield the same amount of nutrients as the estimated dung output of the country in 1966. To get the same amounts of nutrients from respective types of chemical fertilizers they used the following nutrient contents: ammonium sulphate: 20.6 per cent nitrogen; superphosphate: 44.46 per cent phosphoric acid, and muriat of potash: 66.0 per cent potassium oxide. Garg et al. (1971), however, have reported different proportions of nutrients in dung. According to them, *wet dung* contains 0.5 per cent N, 0.2 per cent P_2O_5, and 0.5 per cent K_2O.

Two questions arise now: (*i*) what are the amounts of different nutrients contained in the portion of dung output used as manure in a given year, and (*ii*) how much of each of the different chemical fertilizers would be required to replace the nutrients from the dung-manure? Consider the young cattle stock again. Let d^*_y be the dung output per young cattle per day, α be the dry matter fraction (refer to equation (4.8) and μ the fraction used as manure. Then using the percentages of nutrients from Garg et al. (1971), which is the latest available source, we get the different nutrients from dung output per day per animal as follows

$$N = 0.005\ d^*_y \times \alpha \times \mu \tag{4.15}$$

$$P_2O_5 = 0.002\ d^*_y \times \alpha \times \mu \tag{4.16}$$

$$K_2O = 0.005\ d^*_y \times \alpha \times \mu \tag{4.17}$$

Consider replacement of N as given by equation (4.15). Suppose that an amount χ' of ammonium sulphate will be required to replace it, then

$$0.206\ \chi' = 0.005\ d^*_y \times \alpha \times \mu$$

or, $$\chi' = \frac{0.005}{0.206}\ d^*_y \times \alpha \times \mu \tag{4.18}$$

Equation (4.18) gives the replacement amount of ammonium sulphate for dung manure per animal per day. Multiplying it by 365 and the total number of the young cattle, we get the total replacement amount of ammonium sulphate as

$$\chi = \frac{0.005}{0.206}\ d^*_y \times \alpha \times \mu \times Y \times 365 \tag{4.19}$$

where Y denotes total young cattle population. The general formula for getting the replacement amounts of different chemical fertilizers *vis-à-vis* dung output of the young cattle stock can be written as

$$\chi_{iy} = \frac{\pi_{id}}{\pi_i}\ d^*_y \times \alpha \times \mu \times Y \times 365 \tag{4.20}$$

where i = 1, 2, 3 denotes nutrients, π_{id} denotes fractions of the ith nutrient from dung manure, π_i fraction of ith nutrient in the ith fertilizer, and Y the population of the young cattle stock. For any category of the cattle population, say j (j = 1, 2, ... n'') we can write equations analogous to equation (4.20) regarding the replacement amounts of different fertilizers annually. Thus, we have the general formula

$$\chi_{ij} = \frac{\pi_{id}}{\pi_i}\ d^*_j \times \mu \times C_j \times 365 \tag{4.21}$$

where χ_{ij} stands for replacement amount of ith fertilizer from the jth category of cattle stock, d^*_j for per day dung output per animal of the jth category, and C_j for the jth category of cattle population. The parameters, π_{id}, π_i, α and μ remain unchanged across all categories of cattle.

There are two aspects of χ_{ij} that need to be clearly understood. Its sum over j (j = 1, 2, ..., n'') yields total amounts of different fertilizers of the ith type that will be required to replace the plant nutrients from the dung manure of the cattle stock of the country. Seen from the viewpoint of the livestock production system, those replacement amounts mean that the system obviates the need of those amounts of fertilizers for agricultural production. In other words, those are the amounts of *chemical fertilizers saved* due to the system. The amount of the ith category of fertilizer saved is

$$\chi_i = \sum_{j=1}^{n''} \frac{\pi_{id}}{\pi_i} d^*_j \times \alpha \times \mu \times C_j \times 365, (j = 1, 2, \ldots n''), (i = 1, 2, 3) \quad (4.22)$$

The other aspect is the CO_2 emission prevented that would otherwise occur if dung manure were to be replaced by chemical fertilizers. Probably for a large size country like India the amount of fossil fuel (diesel) used in transportation and distribution of the chemical fertilizers is more than in their manufacture (net of its feed-stock). For the present we focus attention on manufacture related emission alone. Let e_{ico_2} denote CO_2 emission in volume per unit of the ith fertilizer associated with its manufacture. The total emission due to its replacement amount is then

$$E_{ico_2} = e_{ico_2} \chi_i \quad (4.23)$$

The total emission from all the three types of fertilizers is then given as

$$E_{co_2} = \sum_{i=1}^{3} E_{ico_2} \quad (i = 1, 2, 3)$$

Eco_2 can be converted into its methane equivalent. Suppose the relative global warming potential of CH_4: CO_2 is ρ: 1.0, then

$$E_{CH_4} = \frac{1}{\rho} Eco_2 \quad (4.24)$$

where E_{CH_4} is methane equivalent of CO_2 emission prevented (see Section 4.5). Therefore, E_{CH_4} is to be credited to methane emission account of the livestock production system.

4.4 Methane Emission Due to Dung Used as Domestic Fuel

The net CO_2 emission due to small scale, dispersed burning of biomass like dung-cake is supposed to be nil, since the emitted CO_2 is reabsorbed by the plant biomass. However, the emission of trace gases associated with CO_2 emission is to be taken into account (IPCC 1995). As already noted, a portion of the annual dung output in India is converted into dry dung-cake and burned for domestic fuel. In the present section our concern is to determine CH_4 as a trace greenhouse gas of CO_2 emission due to burning of the dung-cake output. Before proceeding, note that CH_4 emission from 'manure management', including the portion converted into dung-cake, has already been accounted for in Section 4.2.2.

Following equation (4.8), α is the dry fraction of dung output of any category of animal, μ is the fraction used as manure as in equation (4.15), and (1-μ) is the fraction of the dung output used for making dung cake. The dung-cake output from the cattle stock in any year is given as

$$D_c = \sum_{j=1}^{n''} (1-\mu) \times d^{*}_{j} \times C_j \times \alpha \times 365, \ (j = 1, 2, ..., n'') \tag{4.25}$$

where d^{*}_{j} is wet dung output per day per cattle of the *j*th category, and C_j its population. The sum over *j* gives us the total of dung-cake output from the cattle stock.

Following IPCC (1995), the amount of CH_4 emission depends upon: (*i*) carbon (mostly CO_2) released from burning of D_c and (*ii*) the ratio of CH_4 to carbon released, namely CH_4 'emission ratio'. The carbon released in turn depends upon two factors: (*i*) carbon fraction of the dung-cake, and (*ii*) the fraction oxidized in the process of burning. Let the carbon fraction of the dung-cake be p and the fraction oxidized be p_o, then the carbon released from burning of D_c is

$$C_f = D_c \times p \times p_o \tag{4.26}$$

And, suppose that emission ratio of CH_4 to CO_2 in the process of burning is e_{CH_4}, then the total CH_4 emission from burning of D_c is obtained by multiplying equation (4.26) by e_{CH_4}. Therefore, total CH_4 emission from D_c is obtained as

$$D_{cCH_4} = e_{CH_4} \times C_f \tag{4.27}$$

Measurements of carbon fractions in different biomass fuels are far from precise, and a wide range of estimates are reported (Delmas and Ahuja 1993). Similarly, there is a very wide range reported for the emission ratio between CH_4 emission and CO_2 released. For instance, whereas Crutzen and Andrea (1990) reported CH_4 to CO_2 emission ratio between 0.007 and 0.013 (mean value 0.01) for all types of biomass fuel, Delmas and Ahuja (1993) reported a different ranges of ratios for different fuels. For dung fuel, however, they reported a point estimate of 0.017, about 59 per cent higher than Crutzen and Andreaes' mean value (0.01). There is, therefore, a great measure of uncertainty surrounding the true values of the parameters involved in the estimation of CH_4 emission from burning of the dung-cake output. In empirical measurement of D_{cCH_4}, one should, therefore, opt for a range rather than a point estimate. And, for this reason debiting of CH_4 emission on this account to the CH_4 emission balance of the livestock production system should be on either–or basis.

4.5 Carbon Dioxide Emission Prevention Due to Animal Energy Use in Agriculture

In addition to methane emission from India's livestock production system, there is, at the same time, prevention of greenhouse gases due to the system. The system annually reproduces and maintains a certain number of working animals (adult cattle and male buffaloes), which are deployed to perform diverse agricultural operations from ploughing of fields to threshing of crop harvest. In agricultural off-seasons they are deployed in rural transport of men and materials. They are a renewable source of energy in contrast to agricultural machinery run on fossil fuel. Work is performed by the animals using muscular energy derived from feed and fodder fed to them. If the working animals were replaced by agricultural machinery, there would occur greenhouse gases emission, of CO_2 in particular from burning of the fossil fuel required to run that machinery. It is this emission that is prevented by the livestock production system of the country.

The problem before us is to determine how much of greenhouse gases emission is prevented annually by the stock of working animals? Besides tractors, there are other machines such as power-tiller, tube-wells, irrigation pumps, threshers, etc., which substitute for or displace working animals. They all use fossil fuel, either diesel oil directly, or coal based thermal electricity. For our present purposes we shall confine attention to tractors only, the ubiquitous machine that substitutes for working animals. Secondly, we shall leave aside emission of trace gases and consider only CO_2 emission from combustion of diesel oil in running a tractor. We shall assume that tractors are alike and the working animals too are alike. Finally, even though a pair of bullocks and a tractor may be used in rural transport, we shall assume that they are held by farmers primarily for performing agricultural operations.

Now, in order to determining CO_2 emission prevented by the stock of working animals we need to know: (*i*) the rate of substitution between tractors and working animals, (*ii*) the amount of diesel oil required for running a tractor to perform the agricultural operations, and (*iii*) the amount of CO_2 emission that will occur from burning a unit of diesel oil. As regards the first it must be considered for the same cultivated area. One may consider net cultivated (sown) area as the common denominator on the ground that farmers invest in and own tractors or working animals keeping their net cultivated area in mind.

It is quite common to take an engineering view of substitution between working animals and tractors. An average bullock is usually rated at 0.4 or 0.5 horsepower. A 25 HP tractor is, therefore, supposed to substitute for or displace at least 50 bullocks. Farm-level studies on tractorization, and analysis of state-level data, do not however, yield such a high rate of substitution. Studies of this nature alone incorporate the economic compulsions of the farmers in doing the substitution. According to such studies, the rate of substitution was in the range of 4 to 10 bullocks per tractor (Mishra and Sharma 1990). Such a wide range is also problematic, if one is looking for an all-India representative figure. Notwithstanding this, the point remains that it is not the engineering but economic rate of substitution that reflects the ground reality. Drawing on the experience of tractorization of agriculture during the last three decades and its effect on the stock of working animals, we propose to determine the economic rate of substitution in the following way. Let a_t and a_{t+n} be the number of working animals per unit of net sown area (say per 100 hectares), at time t and $t + n$ respectively. And, let b_t and b_{t+n} be the corresponding number of tractors. Then, the economic rate of substitution, s, is given by

$$s = (-)\ \frac{a_{t+n} - a_t}{b_{t+n} - b_t} \tag{4.28}$$

The numerator and denominator on the right hand side indicate the change in density of working animals and tractors respectively. The negative sign in front of the right hand side of the equation is due to the fact that the density of working animals is supposed to decrease over the years due to tractorization. While estimating *s* two precautions are necessary. Firstly, *t* and *t+n* should be normal agricultural years. And, secondly, the time period difference between them should be sufficiently large.

If in the given year, the total number of working animals is *A*, then the number of tractors that would be required to displace it is

$$T = \frac{1}{s} \times A \tag{4.29}$$

Given the amount of diesel oil used per tractor during an agricultural year, the total quantity of diesel that would be required to run the substitute number of tractors is obtained as

$$P = p^* \times T \tag{4.30}$$

where p^* is the diesel used per tractor per year. Suppose that σ is CO_2 emission in weight in the process of combustion of a unit of diesel in weight, then the total $\overline{CO_2}$ emission that would occur in the absence of the working animals is

$$\overline{CO_2} = \sigma P \tag{4.31}$$

In the presence of the working animals' stock, *A*, it becomes a negative quantity, being the amount of CO_2 emission actually prevented. It should, therefore, be credited to India's livestock production system. To do that we have to convert it into methane and subtract the latter from total methane emission from enteric fermentation and manure management. Conversion is done in terms of their relative global warming potential (GWP). The measurement of GWP of different greenhouse gases is not quite precise. Though methane admittedly has many times higher GWP compared to CO_2, the reported estimates of GWP of CH_4 in comparison to CO_2 range from as low as 9 to as high as 24.5. The IPCC recommendation for the relative GWP of CH_4 and CO_2 is 21:1. The amount of CH_4 equivalent to $\overline{CO_2}$ then is

$$E'_{CH_4} = \frac{1}{\rho}\overline{CO_2} \qquad (4.32)$$

where ρ is the ratio of GWP of CH_4 to CO_2.

The animal energy use in Indian agriculture may be deemed as a greenhouse gases' mitigation strategy already in place, which is built in the mixed farming system. At a recent greenhouse gases protocol meeting the US claimed credit for its wood and paper waste landfills as carbon sinks like forests. If that claim is true it is not clear why *ex ante* emission prevented by the working animals stock or for that matter any such bio-resource should not be given a credit in India's national inventory of greenhouse gases. Finally, in the present section we have confined attention to CO_2 emission prevented in a given year. However, prevention is over an indefinite period of time. This aspect can be handled in a dynamic version of the model.

4.6 Methane Emission Balance

The balance of methane emission from India's livestock production system is the sum of emissions on account of (*i*) enteric fermentation, (*ii*) 'manure management', and (*iii*) burning of dung-cake minus the sum of CH_4 equivalents of CO_2 emission prevented on account of (*i*) animal energy used in agriculture, and (*ii*) dung used as manure.

In respect of the cattle population, equation (4.14) gives the sum of CH_4 emission from enteric fermentation and manure management. Emission from burning of dung-cake output is given by equation (4.27). By adding equation (4.27) to equation (4.14) we get the total CH_4 emission due to the cattle stock:

$$CCH'_4 = CTE_{CH_4} + D_{cCH_4} \qquad (4.33)$$

On the credit side, CH_4 equivalent of CO_2 emission prevented due to animal energy use in agriculture is given by equation (4.32), and that due to dung used as manure is given by equation (4.24). The sum of these two gives the CH_4 emission prevented as

$$CCH''_4 = E'_{CH_4} + E_{CH_4} \qquad (4.34)$$

The balance of CH_4 emission from the cattle stock is then given as

$$CCH_4 = CCH'_4 - CCH''_4 \qquad (4.35)$$

Proceeding in a similar manner we can arrive at the balance of CH_4

emission from the buffalo stock. By adding it to the balance from the cattle stock we can get the final balance of methane emission from the livestock production system.

4.7 Methane Emission from Dairy Animals

While dealing with methane emission from livestock sources it is quite common to present estimates of emission from dairy animals separately. The idea is to show methane production *vis-à-vis* the efficiency of the dairy production system. For the same level of milk output, if milk-yield per dairy animal is higher, it requires a smaller number of animals to produce it, and CH_4 emission from enteric fermentation per unit of milk output is expected to be lower, even though emission per animal may be higher. Such an efficient system of dairy production is usually recommended as a strategy for reduction of methane emission. What will be the natural resource cost, for instance, land required for producing feed-grains or green fodder for raising the milk yield, will be in order for a country to shift to such an efficient system is generally overlooked. The point to note is that for the same level of milk output if a dairy production system, S_1, causes less CH_4 emission than another system, S_2, but requires more of land than S_2, then one cannot claim that S_1 is *environmentally superior*, hence preferable to S_2 (see Chapter 10, Section 10.5.1). Though the Indian dairy production system, which relies largely on feeding by-products of agriculture requiring no additional land, may cause larger amount of CH_4 emission for the same level of milk production as in a developed country where dairy animals are fed feed-grains requiring land for their production, it cannot be said to be *environmentally inferior* in comparison to the developed country in question.

There is another way of looking at this sort of indeterminacy. Anthropogenic interventions in the natural reproductive process of a dairy cattle population using available scientific means, and driven by efficiency considerations (getting more and more milk per animal) necessarily results in two things: (*i*) a highly *selected* population in which most dairy animals are related to each other; and (*ii*) a breeding efficiency almost equal to one. Both of these inevitable consequences of intervention require some explanation. The first one implies an environmentally damaging effect by reducing the genetic variability of the population. It is reported that in the US

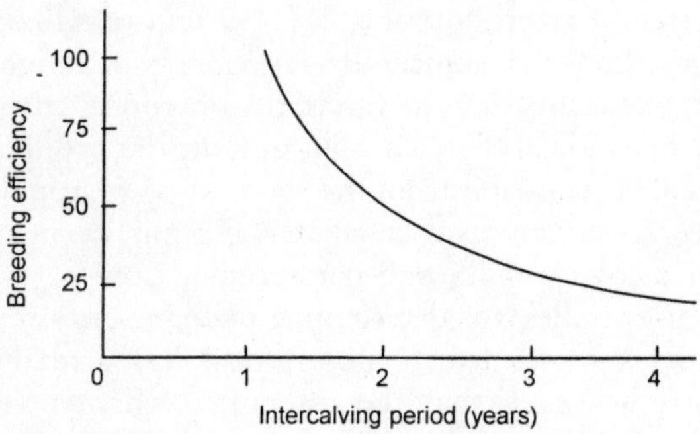

Source: Mishra 1978.

FIG. 4.1: BREEDING EFFICIENCY AND INTERCALVING PERIOD

presently there are only 66 cows, which are unrelated to each other (de Haan et al. 1997).

For understanding the implications of the second consequence, consider Fig. 4.1, which on the basis of a *priori* knowledge postulates an inverse relationship between 'breeding efficiency' of a cow population and the intercalving period. The breeding efficiency is defined as the ratio of cows giving birth to a calf (cows in-milk) during the year to total breeding cows (cows in-milk and dry cows). The curve represents the graph of an equation,

$$e \times k = 1 \tag{4.36}$$

where e denotes breeding efficiency and k denotes the intercalving period, measured in calendar year. When the intercalving period is one year, being the limiting case, assuming 9 months of pregnancy period, and 3 months of waiting before the next conception, e is equal to 1. This implies that each member of the cow population gives birth to a calf any day of the year. All cows are in-milk; there are no dry cows. As k increases, e falls, approaching zero asymptotically for very large values of k. This is, of course, the theoretical position. The real dairy production system will lie somewhere on the left, upper portion of the curve. By highly selective breeding, scientific feeding, and management, and by changing the natural reproductive traits of the cattle population to standardized patterns, the industrial dairy production systems seem to have approached the uppermost limit of e tending to 1.0, virtually eliminating thereby what we call dry cows. It is reported that more than 90 per cent of milk cows in the US give birth to a calf during the year. Such a system may produce less CH_4 per unit of milk output, but it cannot be called environmentally superior if due weightage is given to genetic variability and the variability of reproductive traits.

Be that as it may, given the cow milk output, the breeding cow population, the CH_4 emission from the stock of cows in-milk, and from the dry cows, one can easily calculate CH_4 emission per unit of milk output, per cow in-milk, and per breeding cow per year. In Section 2.2, the procedure for determining methane emission from various categories of the cattle population has already been described. Note, however, that the category of dry cows there includes (*a*) adult females not calved even once, and (*b*) 'others'. For the purposes of the present section the dry cow population is exclusive of these two categories. Following equation (4.3) we can find out CH_4 emission per year per cow in-milk, and per dry cow from enteric fermentation. Similarly, following equation (4.9) we can find out the corresponding CH_4 emission rates from manure management. Let C_m denote cow population in milk, and C_d the population of dry cows, then methane emission from the respective populations during the year is

$$CH_{4m} = (q + q')\, C_m \tag{4.37}$$

and

$$CH_{4d} = (g + g')\, C_d \tag{4.38}$$

where q and q' are the CH_4 emission rates per cow in-milk per year from enteric fermentation and manure management respectively; and g and g' are the corresponding rates per dry cow. The total emission from the breeding cow population is obtained by summing equation (4.37) and (4.38) as

$$CH_{4b} = CH_{4m} + CH_{4d} \tag{4.39}$$

Now, let the total cow milk output during the year be O_c. Then CH_4 emission per unit of milk output is

$$V = \frac{CH_{4m}}{O_c} \tag{4.40}$$

and

$$V' = \frac{CH_{4b}}{O_c} \tag{4.41}$$

where V denotes emission per unit of milk when cows in-milk only are considered, and V' when the total breeding population is considered.

4.8 Saving of Land Due to the Livestock Production System

As pointed out earlier, there are two ways in which India's livestock production system replaces or saves land: (*i*) by using or recycling agricultural by-products as animal feed, and (*ii*) by utilization of a portion of annual dung output for domestic fuel (Section 4.6). If the by-product feeds were to be replaced by feed-grains or cultivated green fodder, the country will require a reallocation of area in favour of these feeds out of the nearly constant net cultivated area of about 143 million hectare. Similarly, if dung-cake output were to be replaced by an alternative biomass fuel, say fuel wood, without causing deforestation, land will be required for fuel wood plantation in order to supply the replacement amount of fuel wood annually. The land saving in either case is a positive environmental contribution of India's livestock production system. For, any activity that displaces a natural resource like land is a nature conserving activity. In the present section our concern is to determine the land area saved due to the livestock system. In the following subsection we deal with the by-product feeds. The dung-cake case is discussed in section 4.8.2.

4.8.1 Land Saving Due to Agricultural By-Products Feed

The problem here can be posed in the following way: If the gross energy intake of the bovine population made available from by-product feed, concentrates, and dry fodder were to be replaced by energy from cultivated green fodder or feed-grains how much of either type of latter feeds will be required. Once the quantity of the alternative feed has been determined, the land that will be required to produce it can be determined. Since some information relating to green fodder production is available in the country, we shall consider cultivated green fodder as the alternative source of energy.

The gross energy intake per day per animal belonging to young cattle stock is given by equation (4.1). The expanded form of the equation is:

$$gei_y = \left(f_{1y} \times d_1 \times e_1\right) + \left(f_{2y} \times d_2 \times e_2\right) + \left(f_{3y} \times d_3 \times e_3\right)$$

where f_{1y}, f_{2y}, f_{3y}, denote concentrates, dry fodder, and green fodder intakes of the animal; d_1, d_2, and d_3 are, respectively, dry matter fractions of the feeds; and e_1, e_2, and e_3, denote energy values per unit feeds of the respective type on dry matter basis. The expressions within braces on the right hand side of the equation give the energy intake from concentrates, dry fodder, and green fodder in that order. The sum of the first two terms gives the total energy intake from the by-products feed.

To simplify matters let us suppose that

$$x_1 = (f_{1y} \times d_1 \times e_1) \tag{4.42}$$

and

$$x_2 = (f_{2y} \times d_2 \times e_2) \tag{4.43}$$

The gross energy intake from by-products can be then written as

$$gei_y^* = x_1 + x_2 \tag{4.44}$$

Now suppose that z_y is the quantity of green fodder on dry matter basis. which will be required to replace gei_y^*. Then, the energy from z_y must equal gei_y^*, such that

$$e_3 z_y = gei_y^* \tag{4.45}$$

where e_3 is the energy (M cal) per unit of z_y. Therefore,

$$z_y = \frac{1}{e_3} \times gei_y^* \tag{4.46}$$

Given the dry matter fraction of green fodder, d_3, we can get green fodder on dry matter basis by dividing equation (4.46) by d_3. Thus,

$$z'_y = \frac{1}{e_3} \times gei_y^* \times \frac{1}{d_3} \tag{4.47}$$

where z'_y stands for the quantity of green fodder that would be required to replace the energy per day per animal from by-products. By multiplying equation (4.47) by 365 we get the green fodder that would be required per year,

$$z''_y = \left[\frac{1}{e_3} \times gei_y^* \times \frac{1}{d_3}\right] \times 365 \tag{4.48}$$

Now we can proceed to find out the land area that would be required to produce z''_y. Let λ be the yield of green fodder per hectare of land, and ℓ_y the area that would be required to produce z''_y, then

$$\ell_y = \frac{z''_y}{\lambda} \tag{4.49}$$

Multiplying equation (4.49) by the young cattle population we get the total area that would be required to produce green fodder in order to replace the energy it receives from the by-products feed,

$$L_y = \frac{z''_y}{\lambda} \times Y \tag{4.50}$$

where *Y* represents young cattle population.

Since we know per day per animal consumption of concentrates and dry fodder for different categories of the cattle population, and their energy contents on dry matter basis we can write equations analogous to equations (4.42)–(4.50) for each category of the cattle population. But to cut it short and focusing only on equations (4.48)–(4.50), the general formula for determining the land saving due to by-products feed fed to the cattle stock can be approached as follows.

Let C_j, ($j = 1, 2, ..., n''$) denote the population of the *j*th category of cattle population and $gei_{c_j p}$ denote the energy intake from by-products feed per day per animal of the *j*th category. Then the green fodder per animal of the *j*th category required to replace its intake of energy from by-products feed is obtained as

$$z''_{c_j} = \left[\frac{1}{e_3} \times gei^*_y \times \frac{1}{d_3}\right] \times 365 \tag{4.51}$$

where subscript c_j denotes the *j*th category of cattle. The land required to produce z''_{c_j} is

$$\ell_{c_j} = \frac{z''_{c_j}}{\lambda} \tag{4.52}$$

Multiplying equation (4.52) by the population of the *j*th category, we get the total annual requirement of land for producing green fodder that would replace the energy intake of the *j*th population from by-products feed,

$$L_{c_j} = \frac{z''_{c_j}}{\lambda} C_j \tag{4.53}$$

C_j being the population of the *j*th category cattle. The total land required to produce green fodder for replacing the energy intake of the cattle stock as a whole from by-products is given as

$$L_{gc} = \sum_{j=1}^{n''} \frac{z''_{c_j}}{\lambda} \times C_j \,, \; (j = 1, 2, ..., n'') \tag{4.54}$$

Proceeding in a similar way we can determine the land area that would be required to produce green fodder to replace the annual energy intake of the buffalo population from by-products feed.

Though our preceding exposition has been made keeping an eye on computer applications of the model, the matter of determining the land area saved due to recycling of agricultural by-products as animal feed is indeed simple. Given the daily feed intakes of by-products feed per animal belonging to different categories of the bovine population, the dry matter fractions of different types of feed, their calorific values, the yield of green fodder per unit of land, and the populations under different categories, it is straightforward to calculate the total land area that would be required to produce green fodder for replacing the by-products feed. And, it is this area that is otherwise saved.

4.8.2 LAND SAVING DUE TO DUNG USED AS DOMESTIC FUEL

In order to determine land saving due to the dung-cake output used as domestic fuel, we shall assume that fuel wood is the alternative biomass source of thermal energy that can and should replace dung-cake. There are justifiable reasons for this assumption. Firstly, like dung-cake, it is a biomass source of energy. The net CO_2 emission from small scale dispersed burning of fuel wood, as in the case of dung-cake, is supposed to be zero. Secondly, like dung-cake, fuel wood is a sustainable source of energy, as it can be supplied from continuously renewable fuel wood plantations.

The dung-cake outputs on dry matter basis from cattle populations is given by equation (4.26). Let us now suppose that ε is the rate of substitution of dry fuel wood for dung-cake (fuel wood: dry dung-cake) in terms of thermal energy. The total quantity of dry fuel wood that would be required to replace the supply (output) of dung-cake in any given year from cattle population D_c [see equation (4.25)], is given by

$$W_t = \varepsilon \times D_c \tag{4.55}$$

Now, further suppose that fuel wood is produced and used within the year, and its yield per hectare on dry matter basis is q^*. Then the total land area that would be required for producing W_t will be

$$L_t = \frac{1}{q^*} \times W_t \tag{4.56}$$

Substituting for W_t from equation (4.55) in equation (4.56), we get

$$L_t = \frac{1}{q^*} \times \varepsilon \times D_c \tag{4.57}$$

The land resource saved due to dung-cake output used as domestic fuel can be computed from equation (4.57), provided we know the values of the parameters q^* and ε and the variable D_c.

4.8.2.1 EFFECT OF GESTATION LAG IN THE PRODUCTION OF FUEL WOOD

The assumption in the preceding section that fuel wood is produced and used within the same year is, however, unrealistic. In fact, it takes more than a year's time to complete the whole process of fuel wood tree plantation, growth and logging of fuel wood trees, drying, and finally using the cut out trees as fuel. For the sake of exposition, suppose that it takes 3 years to complete the whole process, before the bone-dry wood becomes available as fire wood at the end of the third or beginning of the fourth year. This means that whereas the fuel wood made available can replace equivalent quantity of dung-cake only in the fourth year, the necessary land area required for growing and harvesting of fuel wood trees for making the fuel wood available will have to be kept locked up during the preceding three years. This implies that three times as much land would be required if the year's dung-cake output were to be replaced by fuel wood than what is given by equation (4.57). This happens because whereas dung cake is a flow out of a *flow*, fuel wood is a *flow* out of a *stock*.

The stock–flow relationship is illustrated in Table 4.1. This table has been prepared under the following assumptions: (*i*) the gestation lag between planting, harvesting, and supply of dry fuel wood is 3 years, and (*ii*) the fuel wood yield on dry matter basis per hectare is 30 tonnes. It should be clear from the table that to get a flow of 30 tonnes of fuel wood in year t, a total of 3 hectares of land would be required, one hectare each planted during the preceding years, $t-3$, $t-2$ and $t-1$. Similarly, to get the same quantity of fuel wood in year $t+3$, one hectare of land will be required to be planted in the years $t+0$, $t+1$ and $t+2$.

The column sums in Table 4.1 give the total area that will remain occupied by fuel wood plantation in order to make available 30 tonnes of fuel wood each year starting in year t_0.

TABLE 4.1: LAND REQUIRED FOR ANNUAL FUEL WOOD FLOW FROM FUEL WOOD TREE PLANTATION

Items	Time Period							
	t-3	t-2	t-1	t+0	t+1	t+2	t+3	t+4
Land planted by fuel wood	1	1	1					
saplings (hectare)	-	1	1	1				
	-	-	1	1	1			
	-	-	-	1	1	1		
	-	-	-	-	1	1	1	
	-	-	-	-	-	1	1	1
	-	-	-	-	-	-	1	1
	-	-	-	-	-	-	-	1
Total land occupied during any year (hectare)	1	2	3	3	3	3	3	3
Annual flow of fuel wood output (tonnes)	0	0	0	30	30	30	30	30

The essential point to note is that when the gestation lag between planting and harvesting of fuel wood trees is 3 years, 3 hectares of land is required to be kept under plantation to yield 30 tonnes of fuel wood per hectare. If the gestation lag is increased to 4 years, it can be easily seen that 4 hectares of land would be required to give the same annual yield of fuel wood. Through this process of induction we can arrive at the following general proposition. If the gestation lag between planting and harvesting of fuel wood trees is h number of years and q^* is the fuel wood yield per hectare, h hectares of land are required to be kept under plantation constantly for q^* amount of fuel wood to flow each year.

Incorporating the effect of gestation lag in equation (4.56) we get the total land area that would be required to produce fuel wood to replace dung cake,

$$L'_t = h \times \frac{1}{q^*} \times W_t \quad (4.58)$$

Substituting for W_t from equation (4.56), we get

$$L'_t = h \times \frac{1}{q^*} \times \varepsilon \times D_c \quad (4.59)$$

Since D_c is known from equation (4.25), we can compute the value of L'_t if we know the values of h, q^*, and ε. Note that it is this order of land saving that occurs year after year due to dung used as domestic fuel.

4.9 Saving of Chemical Fertilizers Due to Dung Used as Manure

In Section 4.3 while dealing with dung used as manure we focused on CO_2 emission prevention due to the dung used as manure. That is, however, only one of its positive environmental effect. Simultaneously there is the saving of chemical fertilizers themselves. This too needs to be taken into account while assessing the environmental effects of the country's livestock production system. The amounts of different chemical fertilizers saved due to the portion of dung output of the cattle stock used as manure are determined by equation (4.22). *Mutatis mutandis*, the same equation can be used to determine the fertilizer saving due to dung manure from the buffalo stock.

4.10 Concluding Remarks

In the global environmental context, the focus of analysis is usually confined to methane emission from livestock sources. In the model presented above, we have taken a comprehensive environmental view of India's livestock production system. The model is modular in structure. Some of the modules can be linked, for instance the modules relating to greenhouse gases emission to find out the methane emission balance of the system. The computational results of the model are presented in the following chapters. The data relating to the variables and parameters are discussed at appropriate places in the chapters concerned. The results relate to the livestock census year of 1992 since the livestock population data used are from the Census Report of 1992. Once the data from subsequent Censuses become available, the model, with appropriate changes in the values of the variables and parameters, can be used to determine the environmental effects of the livestock production system in the subsequent census year. The model can also be simulated to ascertain the effects of uncertainty regarding parameter values.

5. Methane Emission

5.1 INTRODUCTION

In the literature on global climatic change, methane emission from livestock sources finds an important place. Global methane emission, that is annual uptake to the atmosphere, was estimated in the range of 500 to 550 million tonnes in the 1990s. Livestock sources account for 16 to 18 per cent of the total methane emission. It seems that methane emission from the world's livestock has remained stagnant. Two reasons have been advanced for this: (*i*) rise in the share of industrial livestock production system at the expense of the pastoral and mixed-farming systems, and (*ii*) beef production growth from cattle having become lower than meat from non-ruminant livestock sources, i.e. pigs and poultry (de Haan et al. 1997). The underlying reasons, however, are different. As for the first, methane emission per unit of production in the industrial system is supposed to be less than in the pastoral or mixed-farming systems. As regards the second, in the Western developed countries there has been a shift in consumer preference from beef to poultry and pig meat for health reasons. The fear of 'mad cow disease' in Western Europe may have played a role in this shift.

As noted in Chapter 3, India's ruminant livestock population is the largest in the world. In cattle, India's share is about 16 per cent of the world's total, and, more than half of world's buffaloes have their home in India. Methane emission from this large sized ruminant population is expected to be large. Several estimates of methane emission from India's livestock population have been made in the past (Singh 1997; NPL 1998; USDA 1990; WRI 1990). These estimates are not comparable as they relate to different years, and hence to different livestock populations (see Chapter 1). One thing, however, appears to be common to them all. They all have proceeded by assuming the average body weights of Indian cattle and buffaloes of different age groups and functional categories. Using those body weights they have worked out the *gross energy requirement* of the average animal, applying the nutritional norms for maintenance, growth, and production. From there only a few steps are involved in estimating the total methane emission from the projected or enumerated population.

The actual feed consumption of the animals through which the animals receive their *gross energy intake* (GEI) does not figure at all in the estimates referred to earlier. And, it is the *intake* rather than *requirement* of gross energy that is relevant for estimating methane emission. Requirement of feed in principle may be less, equal to, or more than actual feed intake. So is the case with gross energy *requirement* and *intake*. Note that most studies to date in India have found that the feed, fodder requirement of the livestock population is *less* than availability (assumed to be equal to consumption) (see Chapter 3). If that is true, then the estimates of methane emission based upon gross energy requirements, *ceteris paribus*, will be overestimates.

In the present study we have made a departure by basing our estimate of methane emission on feed consumption, and hence on gross energy intake. The following section deals with emission from enteric fermentation. Section 5.3 is devoted to emission from 'manure management'. Section 5.4 deals with emission of CH_4 as a trace gas of CO_2 emission from the burning of dung cake output. Methane emission from dairy animals and milk production is discussed in Section 5.5.

5.2 Emission from Enteric Fermentation

As already noted earlier, the data on cattle and buffalo population and their distribution by different age, sex, and functional categories have been culled from the *Indian Livestock Census, 1992*. There are problems regarding the census data (Mishra 1970). Not all states and union territories carry out the census on the same reference date or reference period, and even in the same reference year. Of late the problem has become more acute (Mishra 1999), so much so that the 1997 Census data are still being compiled by the Directorate of Economics and Statistics, Ministry of Agriculture, Government of India. Be that as it may, for obtaining all-India figures, census data need to be corrected for the lapses in enumeration mentioned earlier. For want of time and resources we have, however, accepted the 1992 Census data, as they are available in published form.

For estimating methane emission, the detailed classification of cattle and buffaloes given in the livestock census has been merged to give us four classes: (*i*) young stock (male + female); (*ii*) adult females in-milk; (*iii*) dry adult females; and (*iv*) the adult male stock. The stock of dry adult females includes the census categories: (*a*) dry cows, and (*b*) cows not calved even once, and (*c*) 'others'. The stock of adult males is inclusive of the census categories: (*a*) males used for breeding only, (*b*) males used for work only, (*c*) males used for breeding and work both, and (*d*) 'neither used for breeding, nor work'. As pointed out in Chapter 3, adult females (cattle or buffalo) enumerated as 'others' and adult males enumerated as 'neither used for breeding nor work' consist of economically useless animals maintained on scraps of leftover feed. Their inclusion means that in terms of feed intake, hence gross energy intake, they are treated at par with dry animals and functional adult males respectively. Since the gross energy intake of these animals is negligible, we have estimated methane emission with and without these animals. In case of every type of animal, 6 per cent of the gross energy intake is supposed to be the methane loss. The results are presented in Tables 5.2 and 5.3 respectively for any one to see the difference.

The data on animal numbers, per day per animal feed intake on dry matter basis, gross energy intake, and related parameters are given in Table 5.1. Feed intake on dry matter basis has been worked out from the basic per day per animal feed consumption data from Jain and Singh (1990), reproduced in Appendix Table 3.3. The gross energy intake values given in the last column of Table 5.1 are the sums of energy values of the different types of feed on dry matter basis.

The estimate of methane emission from enteric fermentation, excluding the useless animals (Table 5.3), are taken up for interpretation here. Of the total methane emission of 9.3555 million tonnes from the bovine stock in 1992, the share of cattle and buffaloes broadly conform to their shares in the bovine population. Cattle's share in the population is about 71 per cent while its share in the emission is approximately 68 per cent. The share of buffaloes in emission (32 per cent) is marginally higher than its share in the

TABLE 5.1: VARIABLES AND PARAMETER VALUES

Category of population	Population in 1992 (million)	Per day per animal consumption on dry matter basis (kg)			Per day per animal gross energy intake (million Joules)
		Concentrate	Dry fodder	Green fodder	
	(1)	(2)	(3)	(4)	(5)
Cattle:					
Cow in-Milk	31.562	0.246	4.509	1.273	97.26
Dry Cows	32.799	0.073	3.942	0.99	80.22
Cattle adult male	74.486	0.249	4.932	1.475	107.52
Cattle young stock	65.734	0.039	1.949	0.445	38.89
Total cattle	204.584	-	-	-	-
Buffaloes:					
Buffalo in-milk	25.892	0.852	5.769	2.193	144.57
Dry buffaloes	17.914	0.209	4.698	1.635	106.14
Buffalo adult male	8.086	0.157	4.896	1.628	108.14
Buffalo young stock	32.314	0.032	1.836	0.498	37.97
Total buffaloes	84.206	-	-	-	-
Total bovines	288.79	-	-	-	-
Energy (M cal./kg)		4.38	3.69	4.34	-
Dry matter fraction		0.90	0.90	0.25	-

Sources: Column 1 Indian Livestock Census, 1992; columns 2, 3, and 4 Appendix Table 5.1; Energy values from Krishna et al. 1978; Dry matter fractions from Sen et al. 1978.

population (29 per cent) primarily because buffaloes in-milk have the highest gross energy intake per day (approximately 145 M Joules). Within buffaloes, for the same reason whereas the share of in-milk buffaloes in population is 31 per cent, they account for almost half, 49 per cent of the methane emission from enteric fermentation. Within cattle, adult males, most of which are used for work, account for over 49 per cent of the emission. Their share in the cattle population is largest (over 36 per cent), and also their gross energy intake per day (approximately 108 M Joules), more than that of cows in-milk (97 M Joules). The pattern of methane emission from enteric fermentation between cattle and buffaloes and among various categories within cattle and buffaloes broadly follows the relative importance of the functional groups such as buffaloes in-milk, cows in-milk, and adult cattle males.

TABLE 5.2: METHANE EMISSION FROM ENTERIC FERMENTATION BY BOVINES IN INDIA, 1992 (including useless animals)

Categories of population	Population (million)	Percentage to total	CH_4 emission (MT)	Percentage to total
	(1)	(2)	(3)	(4)
Cattle				
Cow in-milk	31.562	15.43	1.21	18.85
Dry cows	32.799	16.03	1.04	16.19
Cow adult male	74.489	36.41	3.16	49.22
Cattle young stock	65.734	32.13	1.01	15.73
Total cattle	204.584	70.84	6.42	67.72
Buffaloes				
Buffaloes in-milk	25.892	30.75	1.48	48.37
Dry buffaloes	17.914	21.27	0.75	24.51
Buffalo adult male	08.086	09.60	0.35	11.44
Buffalo young stock	32.314	38.37	0.48	15.69
Total buffaloes	84.206	29.16	3.06	32.28
Total bovines	288.790	100.00	9.48	100.00

Note: Dry cows include 'not calved even once' and 'others' categories of adult female cattle and buffaloes respectively; adult male includes all the categories as enumerated under adult males of cattle and buffaloes in *Indian Livestock Census, 1992.*

TABLE 5.3: METHANE EMISSION FROM ENTERIC FERMENTATION BY BOVINES, 1992 (excluding useless animals)

Categories of population	Population excluding useless animals (million)	Percentage to total	CH_4 emission (MT)	Percentage to total
	(1)	(2)	(3)	(4)
Cattle				
Cow in-milk	31.562	15.64	1.2100	19.14
Dry cows	31.118	15.42	0.9868	15.61
Cattle adult male	73.448	36.39	3.1160	49.28
Cattle young stock	65.734	32.56	1.0100	15.97
Total cattle	201.862	70.77	6.3230	67.59
Buffaloes				
Buffalo in-milk	25.892	31.06	1.4800	48.80
Dry buffaloes	17.285	20.73	0.7317	24.13
Buffalo adult male	7.873	9.44	0.3408	11.24
Buffalo young stock	32.314	38.76	0.4800	15.83
Total buffaloes	83.364	29.23	3.0325	32.41
Total bovines	285.226	100.00	9.3555	100.00

Note: Population given in column 2 excludes the population of 'neither used for breeding nor for work' from adult male cattle and buffaloes respectively. And 'other' category excluded from dry cow and dry buffaloes.

5.3 Emission from 'Manure Management'

Since methane emission from 'manure management' is very small in comparison to emission from enteric fermentation, we have not cared to exclude useless categories of animals from the population. The unadjusted number of animals in different categories, for estimating CH_4 emission from manure management are the same as given in column (1) of Table 5.1. The relevant parameter values are presented in Table 5.4. The dung evacuation rates, i.e. dung output per animal per day, are estimated by us (see the note in Appendix: A Note on the Estimation of Dung Evacuation Rate). According to Garg et al. (1971), the moisture content of fresh dung ranges between 75 and 85 per cent. In order not to err on the side of underestimation, we have adopted 75 per cent as moisture content. Based on chemical analysis of dung on dry matter basis, the ash content is reported to be 11.0 per cent (Ranjhan 1998). The figures on volatile solids (VS) in column (5) of Table 5.4 are dung outputs on dry matter basis net

TABLE 5.4: PARAMETER VALUES

Categories of population	Evacuation rate (wet) kg/head/ day	Moisture content (% of wet)	Dry dung (kg/ head/ day)	Ash fraction (%)	Volatile solids (Vs) kg/ head/ day	Maximum methane emission potential (Bo)	Methane conversion factor (%)
	(1)	(2)	(3)	(4)	(5)	(6)	(7)
Cattle							
Cow in-milk	7.74	75	1.94	11	1.73	0.10	2.00
Dry cows	5.97	75	1.49	11	1.33	0.10	2.00
Cattle adult male	7.86	75	1.96	11	1.74	0.10	2.00
Cattle young stock	4.39	75	1.09	11	0.97	0.10	2.00
Buffaloes							
Buffalo in-milk	9.69	75	2.42	11	2.15	0.10	2.00
Dry buffaloes	8.29	75	2.07	11	1.84	0.10	2.00
Buffalo adult male	5.73	75	1.43	11	1.27	0.10	2.00
Buffalo young stock	3.35	75	0.84	11	0.75	0.10	2.00

Sources: Column 1: see Appendix: A Note on the Estimation of Dung Evacuation Rate; column 2: Garg et al. (1971); column 4: Ranjhan (1998); columns 5, 6, and 7: IPCC (1995).

of their ash content. We do not have Indian data on the 'maximum methane emission potential' (Bo) of the volatile solids (VS). Therefore, we have adopted the IPCC's 'default rates' as given in column (6) of Table 5.4. The 'methane conversion factor' (MCF), i.e. the part of the potential actually realized (column 7), is again adopted from the IPCC. As we have noted in the preceding chapter, the IPCC has provided different MCF values for different 'manure management systems'. We have assumed that all dung output in India is handled in solid form, and in aerobic conditions. The MCF value given in column (7) of Table 5.4 relates to this 'system' of manure management. As for the CH_4 emission from burning of dung for fuel, we have treated it separately in the following section. The emission of CH_4 from this is *additive* to emission from 'manure management'.

The output of methane according to various categories of the bovine population from 'manure management' is presented in Table 5.5. The total emission works out to 0.202 million tonnes from the dung output of the bovine stock in 1992. The shares of cattle and buffaloes in the total methane emission are almost equal to their respective shares in the bovine population. Specifically, while the share of cattle is 70 per cent, that of buffaloes is 30 per cent of the total. Within cattle, as in the case of emission from enteric fermentation, adult males account for the largest share, 44 per cent of the total. Similarly, within buffaloes, adult females in-milk

TABLE 5.5: METHANE EMISSION FROM MANURE MANAGEMENT IN INDIA, 1992

Categories of population	Population (in million)	Percentage to total	CH_4 emission (MT)	Percentage to total
Cattle				
Cow in-milk	31.562	15.43	0.027	19.01
Dry cows	32.799	16.03	0.021	14.79
Cattle adult male	74.489	36.41	0.063	44.37
Cattle young stock	65.734	32.13	0.031	21.83
Total cattle	204.580	70.84	0.142	70.29
Buffaloes				
Buffalo in-milk	25.892	30.75	0.027	45.00
Dry buffaloes	17.914	21.27	0.016	26.67
Buffalo adult male	8.086	9.60	0.005	8.33
Buffalo young stock	32.314	38.37	0.012	20.00
Total buffaloes	84.206	29.16	0.060	29.70
Total bovines	288.790	100.00	0.202	100.00

account for 45 per cent of the total. The breedable cows within cattle [(in-milk + dry including not calved even once + 'others')] account for about 44 per cent of the total emission. Quite in contrast, the breedable buffaloes account for 72 per cent of methane emission from the dung output of the buffalo population. This is because per day feed intake of both in-milk and dry buffaloes, and, therefore, their per day dung output, is relatively larger in comparison to other categories of animals.

5.4 Emission from Burning of Dung Cake

The parameter values required for the estimation of methane released as a trace gas in the process of burning of dung cakes are given in Table 5.6. The dung evacuation rates and populations of different categories of animals are the same as in Tables 5.4 and 5.5 respectively. According to the National Income Accounts Division of the Central Statistical Organisation, 40 per cent of the dung output in the country is used for making dry dung-cake, and the remaining is used as manure. The values of the fraction of dung-cake oxidized and the fraction of carbon (mostly CO_2) released are taken from the IPCC (1995). These 'default values' have been adopted, as we are not aware of any national estimates of these parameters. The most uncertain parameter value is that of the ratio of CH_4 emitted to carbon released. Whereas Crutzen and Andreae (1990) have reported a value of 1.0 per cent, Delmas and Ahuja

TABLE 5.6: PARAMETER VALUES

Particulars	Values
Moisture fraction in wet dung	0.75
Fraction of dung used as fuel	0.40
Ratio of CH_4 to carbon released	0.01
Fraction oxidized	0.85
Carbon fraction released	0.36

Sources: Wet to dry ratio of dung: Garg et al. (1971).
Fraction of dung used as fuel, obtained through personal communication from Dr Kulshreshtha, DDG, Central Statistical Organisation (CSO), National Accounts Division.
Ratio of CH_4 to carbon released: Crutzen and Andreae (1990) and IPCC (1995).
Fraction oxidized: IPCC (1995).
Fraction of Carbon released: IPCC (1995).

TABLE 5.7: METHANE EMISSION FROM BURNING OF DRY DUNG-CAKE AS DOMESTIC FUEL, 1992

Particulars	Values (in MT)
Total production of wet dung by total bovines	682.19
Dung output on dry matter basis	170.58
Dung used as domestic fuel in the form of dry dung-cake	68.23
Carbon released from burning of the dung-cake	20.87
Emission of CH_4	0.21

(1993) have reported a value of 1.7 per cent. Other things remaining the same, applying the latter ratio would yield 70 per cent larger CH_4 emission from the same amount of dung-cake. The IPCC (1995) has recommended the latter ratio as the 'default value' for preparing a national inventory of greenhouse gases without giving reasons for this. One is, therefore, free to choose one or the other value of the parameter. We have chosen Crutzen and Andreae's ratio, since we will not like to inflate CH_4 emission from dung-cake burning. There is an added reason for this. Methane emission from the 'manure management system' in the preceding section covers all dung output, including the portion used for making dung-cake.

The results are presented in Table 5.7. The total wet dung output from the bovine stock in 1992 works out to over 682 million tonnes. Out of this, on dry matter basis, 68 million tonnes is the dung cake output burnt as domestic fuel. The carbon released from its burning works out to about 21 million tonnes. One per cent of this, that is 0.21 million tonnes, is the CH_4 emission as a trace gas of CO_2.

5.5 Methane Emission from Dairy Animals and Per Unit of Milk Output

In the Indian dairy production system, buffaloes play a relatively larger role than cows. As pointed out in the preceding chapter, for over several millennia cows in India have been reared and evolved more for producing and reproducing working bullocks than for producing milk. That apart, in any year adult female buffaloes as well as cows that have come in production, that is in the breeding stage (calved at least once), consist of two groups: (*i*) in-milk, and (*ii*) dry. Milk production during the year is from females in-milk, while dry females are in the period of gestation before the next calving. Methane emission from enteric fermentation from females in-milk is directly imputable to milk production during the year. As regards methane emission from dry females, one may take different views. One view could be that they being maintained for milk production, CH_4 emission from them should be imputed to milk production. An alternative view could be that they being in gestation period, preparatory to pregnancy plus the period of pregnancy, CH_4 emission from them should not be imputed to milk production. Rather, it should be imputed to the animals' harmonal change conducive to pregnancy, and the growth of the foetus in the womb. It is true that these processes require gross energy intake, and the loss of energy in the form of methane during the course of these processes is not accounted separately as it ought to be done. This happens simply because the standard is set by the dairy system of the developed Western countries in which anthropogenic interventions in the reproductive processes have virtually eliminated the dry cows, including due to culling of cows in-milk and sending them to slaughter house if they fail to meet the preset norm of milk yield.

Be that as it may, in the present section we have worked out CH_4 emission per unit of milk produced from females in-milk as well as from total breeding females (in-milk + dry). Emission not only from enteric fermentation but also from manure management is included in the exercise. It should be noted that the population of dry females, cattle as well as buffaloes, is exclusive of the economically useless category called 'others' and also adult females (> 3 years of age), 'not calved even once'. The results of the exercise are presented in Table 5.8.

It is interesting to compare CH_4 emission from cows in-milk and buffaloes in-milk. Whereas emission per kg of buffalo milk (49 gm per kg) is not significantly different from emission per kg of cow milk (51 gm per kg), emission per buffalo in-milk per year is very much higher (58 kg) than emission per cow in-milk (39 kg). However, when dry animals are added, the picture changes. Emission per kg of buffalo milk (69 gm) turns out to be significantly lower than emission per kg of cow milk (86 gm). None the less, emission per breeding buffalo per year still remains very much higher (53 kg) than emission per breeding cow (36 kg).

Clearly, on the basis of these two criteria we cannot say that the buffalo based dairy production system is more efficient than the cow based system, since the criteria values point in the opposite directions. Efficiency of any production system is a measure of the relation between the input to the system and its output. A milch

TABLE 5.8: METHANE EMISSION PER UNIT OF MILK PRODUCTION, 1992

Categories of population	Population (in million)	Methane emission (MT)	Milk production (MT)	Gross energy intake (MJ) per animal per day	Milk yield per animal per day (kg)	Milk yield per unit[+] of energy intake (gm)	CH_4 emission per kg of milk produced (kg)	CH_4 emission per animal per year (in tonnes)
	(1)	(2)	(3)	(4)	(5)	(6)	(7)	(8)
Cattle								
Cow in-milk	31.562	1.237	24.338	97.26	2.11	21	0.051	0.039
Dry cows	26.231	0.8496	-	80.22	-	-	-	0.032
Total breeding cows	57.793	2.0866	24.338	89.51*	1.15	13	0.086	0.036
Buffaloes								
Buffalo in-milk	25.892	1.507	31.044	144.57	3.28	22.68	0.049	0.058
Dry buffalo	14.383	0.6222	-	106.14	-		-	0.043
Total breeding buffaloes	40.275	2.1292	31.044	130.83*	2.11	16.13	0.069	0.053
Bovines								
Bovines in-milk	57.454	2.744	55.382	-	-	-	0.05	0.048
Dry bovines	40.614	1.4718	-	-	-	-	-	0.036
Total breeding bovines	98.068	4.2158	55.382	-	-	-	0.076	0.043

Notes: Methane emission (col. 2) includes emission from enteric fermentation and manure management.
* Weighted mean of the gross energy intake of animal in-milk and dry animal.
+ One million Joules.

Sources: *Indian Livestock Census, 1992* for population;
Basic Animal Husbandry Statistics, 1997 for Milk Production;
Methane emission our estimate.

animal is not an input to the system, but a means for transforming input to the system into output. It is gross energy intake of the animals which in fact is the input to the system. Both milk and methane are outputs. From this angle we can set the efficiency criteria as follows. If a system produces more milk per unit of gross energy intake and produces less CH_4 emission per unit of milk produced, it can be said to be more efficient than another system which produces less milk per unit gross energy intake and produces more CH_4 emission per unit of milk produced.

According to this criteria, the buffalo based dairy production system stands out to be more efficient than the cow based system. Per million Joules of energy, a buffalo in-milk produces marginally more milk (23 gm) compared to a cow in-milk (21 gm). At the same time it produces less CH_4 per unit of milk produced (49 gm) compared to its cow counterpart (51 gm). The same position holds when we consider the breeding stocks as a whole. Milk produced per million Joules of energy per breeding buffalo works out to be larger (16 gm) compared to per breeding cow (13 gm). And CH_4 emission per unit of milk produced from a breeding buffalo works out to be lower (69 gm) than from a breeding cow (86 gm). Therefore, the buffalo based dairy production system turns out to be more efficient than the cow based dairy system.

It may legitimately be argued that instead of gross energy intake (GEI), total digestible nutrients (TDN) consumed is a more relevant variable on which milk production depends. We have carried out an exercise with TDN base too (Appendix Table 5.2). The results further confirm higher efficiency of the buffalo based milk production system, rather more substantially in terms of milk produced per unit of TDN consumed. Specifically, milk produced by buffaloes in-milk per kg of TDN is larger (625 gm) than milk produced by cows in-milk (604 gm), while CH_4 emission per kg of milk produced by the former is lower (49 gm) than by the latter (51 gm). A similar position holds when we consider that produced per kg of TDN consumed by breeding buffaloes is larger (424 gm) as against milk produced by breeding cows (367 gm). And, emission of CH_4 per kg of milk produced by breeding buffaloes is lower (69 gm) than by breeding cows (86 gm). Finally, in case one draws a policy inference from this against the cow production system, it is important to remind the reader that there remains an element of incomparability between the two systems. The cow production system, as noted earlier, has been primarily geared to the production and reproduction of bullocks over the ages, a function for which the buffalo production system has been found to be unsuitable.

5.6 Summary and Conclusions

The total methane emission from the livestock production system in 1992 works out to 9.767 million tonnes. Emission from 'manure management' (0.202 million tonnes) and from burning of dung-cake as domestic fuel are insignificant in comparison to emission from enteric fermentation (9.355 million tonnes). A notable point about the estimate of CH_4 emission from enteric fermentation is that it is not based on the IPCC 'default rates'. Nor is it based on the gross energy *requirement* of animals worked out using assumed average body weight and nutritional norms for maintenance and production. It is based upon gross energy *intake* derived from feed consumption rates estimated from primary data collected through sample surveys. An estimate which involves fewer assumptions and more of survey based primary data is supposed to be more realistic.

In the total CH_4 emission from enteric fermentation, the share of cattle (68 per cent) is marginally lower than its share in the bovine population (71 per cent). The share of buffaloes (32 per cent) is marginally higher than its share in the population (29 per cent). Interestingly whereas within buffaloes, in-milk buffaloes account for close to half of CH_4 emission from enteric fermentation, within cattle, adult males, most of them being working bullocks, account for about half of the emission.

From the point of view of methane emission reduction it is usual to focus on dairy animals. In terms of a set of two efficiency criteria, CH_4 emission per unit of milk produced and milk produced per unit of gross energy intake, the buffalo based dairy production system turns out to be more efficient than the cow based system. The buffalo based dairy production system produces more milk per unit of gross energy intake and less CH_4 emission per unit of milk produced than the cow based dairy production system. The same relative efficiency position is confirmed when, instead of gross energy intake, we consider the issue with TDN consumed as the base. However, from a policy perspective it should be kept in mind that the cow based production system's primary function, unlike that of the buffalo based system, is to produce and reproduce working bullocks.

APPENDIX TABLE 5.1: FEED CONSUMPTION ON DRY MATTER BASIS (in kg/animal/day)

S. No.	Categories	Concentrate	Dry	Green	Total
1	Cow in-milk	0.246	4.509	1.273	6.028
2	Dry cow	0.073	3.942	0.99	5.005
3	Adult male cattle	0.249	4.932	1.475	6.656
4	Cattle young stock	0.039	1.949	0.445	2.433
5	Buffalo in-milk	0.852	5.769	2.193	8.814
6	Dry buffaloes	0.209	4.698	1.635	6.542
7	Buffalo adult male	0.157	4.896	1.628	6.601
8	Buffalo young stock	0.032	1.836	0.498	2.366

Sources: Consumption rates of feed is taken from Jain and Singh (1990); Dry matter in terms of per cent of green are taken from Ranjhan (1998).

APPENDIX TABLE 5.2: MILK YIELD PER UNIT OF TDN CONSUMED BY INDIAN CATTLE AND BUFFALO

S. No.	Categories of animal	Population (million)	Milk yield per animal	TDN consumed per animal per day (kg)	Milk yield per kg of TDN consumed (gm)	CH_4 Emission per kg of milk produced
A	Cattle					
1	Cow in-milk	31.562	2.11	3.49	604	0.051
2	Dry cow	26.231	-	2.69	-	-
3	Total breeding cow	57.793	1.15	3.13	367	0.086
B	Buffalo					
1	Buffalo in-milk	25.892	3.28	5.24	625	0.049
2	Dry buffalo	14.383	-	4.49	-	-
3	Total breeding buffalo	40.275	2.11	4.97	424	0.069

Sources: Indian Livestock Census, 1992 for population; Basic Animal Husbandry Statistics, 1997 milk for yield per animal; Jain and Singh (1990) for TDN consumed per animal per day. *Weighted mean of TDN intake of cow in-milk, dry cow, buffalo in-milk, and dry buffalo.

6. Carbon Dioxide Emission Prevented Due to Animal Energy Use in Agriculture

6.1 Introduction

By the mid-1960s, growth in net cultivated area at the all-India level virtually ceased and since then the area has remained stagnant at between 140 and 143 million hectares. Growth in agricultural production since the mid-1960s has been achieved through growth in crop yields and in gross cropped area. Until the mid-1960s, agricultural production in the country was almost wholly dependent on animal draught power. Then, in the wake of the Green Revolution, mechanical draught power began to be introduced as a component of the New Agricultural Strategy. A set of two public policies promoted agricultural mechanization: (*i*) soft credit and subsidy for purchase of tractors and other machinery, and (*ii*) a cattle breeding policy which exclusively focused on crossbreeding of indigenous cows with exotic breeds aimed at maximizing milk production, completely ignoring the fact that indigenous cattle for over thousands of years had been reared and evolved for producing and reproducing bullocks required for agriculture. From a low base the tractor stock grew rapidly. Between 1972 and 1992, the number of tractors increased from 14,000 to 1.22 million. Since the net cultivated area remained stagnant, the density of tractors per unit area too increased rapidly. Its effect could be none other than a decline in the number of working animals per unit area. Mishra and Sharma (1990) had shown that there was for the first time a marginal decline in the total number of working animals between 1977 and 1982. The decline became substantial between 1982 and 1992 from over 80 to 70 million (see Appendix Table 6.1).

No one realized, nor perhaps there is realization even today, that this process has a serious environmental implication. Working animal stock is a renewable, sustainable bio-source of energy. The stock derives this energy largely from agricultural by-products feed, the social opportunity cost of which is virtually zero. Its replacement by machine causes depletion of non-renewable fossil fuel resources here or elsewhere in the world. In addition, burning of the fossil fuel in running the machinery causes greenhouse gases emission, CO_2 and the trace gases.

In the present chapter we have made an estimate of CO_2 emission that is prevented by the stock of working animals, or to put it differently, CO_2 emission that will occur if the working animals' stock were to be replaced by tractors. To the extent that other machinery like irrigation pump sets are excluded from consideration, the estimate we have presented in Section 6.3 is an underestimate. As we have seen in Chapter 4, the key parameter involved in the exercise is the economic substitution rate between working animals and tractors. The following section is devoted to a discussion on this.

6.2 Replacement (Substitution) Rate Between Working Animals and Tractors

The problem can be approached at two levels: (*i*) at the micro farm level, and (*ii*) at the macro state and all-India level. Farm-level studies have to be based necessarily on farms which have been tractorized. And, these are relatively large sized farms. Their results, though very useful as indicators of ground reality, cannot be generalized to the macro state or all-India level because the structure of farm holdings at these levels is dominated by small and marginal sized farms. At the macro level, the structure of farm size holdings, apart from other factors, is a key determinant of the replacement of working animals by tractors.

Farm level studies carried out during the 1970s and 1980s in the western region, in Punjab, Haryana and western Uttar Pradesh which were the leaders in tractorization, reported replacement rates of three to four bullocks per tractor (Binswanger 1978; Sharma 1987; Mishra and Sharma 1990). At the macro level of Haryana and western Uttar Pradesh using the changes in the stocks of working animals and tractors between 1972 and 1982, Mishra and Sharma (1990) reported a replacement rate of four animals per tractor. This macro level result apparently, by and large, confirmed the earlier farm level findings.

As the 1992 livestock census data because available, we ourselves carried out a number of exercises at the state level for a few selected states. Using 1982 data, Mishra and Sharma (1990) have classified all the major states into three groups according to the number of working animals per 100 hectares of net sown area. The groups are: low density states (25–45 animals), medium density states (55–80 animals), and high density states (85–150 working animals). We chose two states from each group: Haryana and Punjab as low density states, Andhra Pradesh and Tamil Nadu as medium density states, and Uttar Pradesh and West Bengal as high density states. The density of working animal's stock is supposed to reflect the differences in landholdings structure and cropping pattern. And, these differences are expected to have a bearing on the replacement rate of working animals by tractors. For working out economic substitution rates three livestock census years, 1972, 1982, and 1992, sufficiently distant from each other were chosen for the exercise. For the selected states, the basic data on net sown area, the number of working animals, tractors, and their densities per 100 hectares of net sown area are given in Appendix Table 6.1. The number of working animals includes (*i*) cattle males (> 3 years) used for work only; (*ii*) cattle males used for work as well as breeding; (*iii*) buffalo males (> 3 years) used for work only; and (*iv*) buffalo males used for work as well as breeding. Before the 1982 Census, small numbers of cattle and buffalo females were also reported as used for work. Since they are not reported from 1982 onwards, we have excluded them for the sake of its comparability with subsequent censuses. The 1982 figures are from Mishra and Sharma (1990), not only for the states but also for the country as a whole. In the 1982 Census, a mindless change was made in the functional classification of adult cattle males, destroying its comparability with previous censuses and also the subsequent ones. The old classification was restored in the 1987 Census when the authorities were made to realize their mistake. In 1982, adult cattle males were enumerated largely under two categories: (*i*) castrated and (*ii*) uncastrated. Mishra and Sharma (1990) have distributed the figures reported under these categories into comparable functional categories and, thus, obtained the number of working animals from the adult male cattle stock. We have, therefore, accepted their figures for 1982. But the fact remains that these are estimates, not census enumerated figures. They could be an over or underestimate. To that extent our estimates of replacement rates involving the 1982 figures will be affected. As for the 1972 and 1992 figures on working animals, these are from the respective livestock census reports.

The results of our exercises for the selected states are presented in Table 6.1. Let us examine the rates of substitution over the ten-year periods, 1972 to 1982 and 1982 to 1992. In the low working animal density states, Punjab and Haryana, the substitution rates turn out to be typically low, not more than one animal in Punjab and not more than three animals per tractor in Haryana in the first decade, 1972–82, which was the high time of Green Revolution in these states. The net sown area increased in both states. The density of working animals in Punjab hardly changed, though there was a marginal fall in Haryana. The reason seems to be the increased *flow* demand of draught power from both the *stocks* due to increased cropping intensity so much so that increase in tractors' density caused only a marginal fall in working animal's density (Haryana) or hardly a fall at all (Punjab). In the following decade, 1982–92, net sown area in Punjab decreased, while tractor density increased five-fold between 1972 and 1992. The net effect was a larger substitution rate, six working animals per tractor. During the decade 1982–92, Haryana presents a reversal of the situation in the 1970s. The substitution rate over 1982–92, works out to no more than one working animal per tractor in Haryana.

In the medium density states, Andhra Pradesh and Tamil Nadu, we come across no set pattern of change and some incredible numbers like a substitution rate of 93 working animals per tractor between 1982 and 1992 but no substitution at all (the rate is positive) in the preceding decade, 1972–82, in Tamil Nadu. In Andhra Pradesh, on the other hand, there is a quite high rate of substitution, 33 working animals per tractor between 1972 and 1982, but no substitution at all between 1982 and 1992. If our basic data are right then the explanation of such pattern of substitution rate reversals and incredibly large differences (from zero to 93 and from 33 to zero per tractor) require a deeper study, looking into changes in the landholding structure, cropping pattern, cropping intensity, and land-use pattern, perhaps even at the district level. For instance, in the case of Tamil Nadu, a substitution rate of 93 animals per tractor between 1982 and 1992 is certainly not due to tractorization. It could be due to transfers of large area to plantations and non-agricultural uses, as there was a large fall in the number of working animals as well as in the net sown area in Tamil Nadu during the period under consideration.

In the high density states, West Bengal is an example of a state in which there was no substitution effect at all. In 1972, West Bengal

TABLE 6.1: RATE OF SUBSTITUTION BETWEEN DRAUGHT ANIMALS AND TRACTORS IN SELECTED STATES

States	In 1982 over 1972			In 1992 over 1982			In 1992 over 1972		
	Change in X1	Change in X2	Rate of Substitution	Change in X1	Change in X2	Rate of Substitution	Change in X1	Change in X2	Rate of Substitution
Low Density									
Punjab	-0.74	1.53	-0.48	-15.14	2.57	-5.89	-15.88	4.10	-3.87
Haryana	-3.05	1.21	-2.52	-2.2	2.78	-0.79	-2.25	3.99	-1.32
Medium Density									
Andhra Pradesh	-4.56	0.14	-32.57	9.86	0.35	28.17	5.30	0.49	10.81
Tamil Nadu	1.51	0.19	7.95	-22.36	0.24	-93.17	-20.85	0.43	-48.49
High Density									
Uttar Pradesh	5.98	0.76	7.86	-14.87	1.18	-12.6	-8.89	1.94	-4.58
West Bengal	16.27	-	-	0.92	0.08	11.5	17.19	-	-

Notes: X1 is the density of working bovines per 100 hectare of net sown area;
X2 is the density of tractors per 100 hectare of net sown area;
There were no agricultural tractors reported for West Bengal in 1972.

Source: Statistical appendix Table 6.1.

did not report any agricultural tractor. In 1982 and 1992 it did report the presence of such tractors, at 2000 and 6500, respectively. During the period the number of working animals increased substantially, and the net sown area marginally declined. The net effect was the increase in the density of working animals as well as of tractors. That there was no substitution effect seems to be due to a change in the landholding structure in favour of small and marginal farmers and increase in the cropping intensity.

Uttar Pradesh, as a high density state, presents a consistent pattern of substitution effect. Between 1972 and 1982, the number of working animals, the number of tractors, and the net sown area all increased in the state. The net effect was no substitution effect at all. In the following decade 1982–92, the number of working animals decreased very substantially, number of tractors increased, and the net sown area marginally declined. As a result, the rate of substitution in this period works out to 13 working animals per tractor.

We also worked out for the selected states substitution rates over the long period of 20 years, that is between 1972 and 1992. Tamil Nadu again showed the incredible rate of 49 working animals per tractor. In Andhra Pradesh, the substitution rate was zero. Punjab showed a substitution rate of four animals per tractor; Haryana at most two animals, and Uttar Pradesh five animals per tractor. Leaving aside the peculiar cases of Andhra Pradesh and Tamil Nadu, the substitution rates over the long period of 20 years range between two and five working animals per tractor. These results confirm the old micro level findings of three to four working animals per tractor as the substitution rate. Yet these rates appear to be too low if one were looking for a rate representative of India as a whole. Notably, these rates are only from low-density states, including the rate for Uttar Pradesh since much of the tractorization there occurred in the low-density western region.

Finally, we worked out the substitution rate at the all-India level. The relevant data and the substitution rates over the period 1972 to 1992 are presented in Table 6.2. During both decades the density of working animals per 100 hectare declined, though only very marginally during 1982–92. Tractor density rapidly increased by more than 300 per cent during the first decade, and again by more than 200 per cent during the second decade. The net sown area hardly increased. The results of the exercise are presented in Table 6.3.

TABLE 6.2: DENSITY OF WORKING ANIMALS AND TRACTORS IN INDIA, 1972–92

Years	Working Animals (million)	Density* of working animals	Tractors (million)	Density* of tractors	Net sown area (million hectare)
1972	80.22	58.52	0.1482	0.108	137.099
1982	80.40	56.71	0.5188	0.365	141.77
1992	70.58	49.54	1.22	0.856	142.5

Note: *Per 100 hectare of net sown area.

TABLE 6.3: RATE OF SUBSTITUTION BETWEEN WORKING ANIMALS AND TRACTORS AT ALL-INDIA LEVEL

Period	Change in density* of working animals	Change in density* of tractors	Rate of substitution col. (1)/col. (2)
	(1)	(2)	(3)
1982 over 1972	-1.81	0.257	-7.04
1992 over 1982	-7.17	0.491	-14.6
1992 over 1972	-8.98	0.748	-12.01

Note: *Per 100 hectare of net sown area.
Source: Appendix Table 6.1.

The rate of substitution over the first decade, 1972–82, works out to 7 working animals per tractor; over the second decade, 1982–92, 15 animals per tractor; and, over the whole period, 1972–92, 12 animals per tractor. For working out the CO_2 emission prevented due to the stock of working animals in 1992 we have accepted 12 animals per tractor as the realistic rate of substitution.It incorporates the complexity of the whole lot of institutional and technological changes that may have occurred in Indian agriculture over this long period of 20 years. It is close to the average of the substitution rates respectively over its sub-periods, 1972–82 and 1982–92. It may also be worthwhile to note that de Haan et al. (1997) have implied a substitution rate of 10 draught animals per tractor when they say that 30 million tractors would be required to replace 300 million draught animals currently used in agriculture worldwide.

6.3 Parameter Values

In addition to the substitution rate of working animals per tractor, the relevant parameter values for the estimation of CO_2 emission prevented by the stock of working animals in 1992 are given in

TABLE 6.4: PARAMETER VALUES

S. No.	Items	Values
1	Consumption of diesel per tractor (tonnes/year)	3.25
2	Carbon fraction of diesel	0.8752
3	Fraction oxidized	0.99
4	Conversion factor from carbon released to carbon dioxide	0.3666

Sources: Mishra and Sharma (1990) for item no. 1; IPCC (1995) for items 2, 3 and 4.

Table 6.4. The consumption of diesel per tractor per year has been taken from Mishra and Sharma (1990). They worked out the rate from the details of agricultural operations in which tractors are used as found in the Manuals on Agricultural Practices of the governments of Punjab and Haryana. The values of the carbon fraction of diesel and the fraction oxidized are from IPCC (1995), as also the conversion factor from carbon released on burning of diesel to carbon dioxide emitted. Incidentally, it is interesting to note that while dung-cake as biomass fuel has a carbon content of 36–40 per cent only, diesel's carbon content is as high as 88 per cent. Similarly, whereas diesel's oxidized fraction is as high as 99 per cent, dung-cake's oxidized fraction is 85 per cent (see Chapter 5, Section 5.4).

6.4 Estimate of CO_2 Emission Prevented

The results of the exercise using the parameter values given in Tables 6.3 and 6.4 are given in Table 6.5. At the substitution rate of 12 working animals per tractor, approximately six million tractors would be required to substitute about 71 million working animals. To operate the tractor stock will require 19.5 million tonnes diesel oil annually. Combustion of this amount of diesel oil will emit over 6 million tonnes carbon dioxide. It is this amount of carbon dioxide emission that is prevented annually by animal energy use in Indian agriculture.

TABLE 6.5: CARBON DIOXIDE EMISSION PREVENTED

Particulars	Values
Numbers of tractors required to replace existing stock of working animals (million)	5.92
Consumption of diesel by required no. of tractors (MT)	19.5
Total carbon released from burning (MT)	16.89
Emission of carbon dioxide prevented (MT)	6.19

6.5 Concluding Remarks

To conclude, the environmental value of animal energy use in Indian agriculture, especially by tens of million of small and marginal farmers, has hardly been realized. This is a renewable and sustainable form of bio-energy. The working animals' population derives it largely from by-products of agriculture whose social opportunity cost is zero. Its alternative fossil fuel energy like diesel is unsustainable and costly, and besides, its burning causes greenhouse gases emission, in particular CO_2 emission.

The agricultural engineers and energy experts who seek equivalence between working animals and tractors in terms of horsepower make the mistake of assuming their normative equivalence to be the same as the economic rate of substitution between the two. The real rate of substitution, on the contrary, is determined by a host of economic factors, such as land use pattern, farm-size, landholdings structure, cropping patterns, etc. Micro farm level studies have found that only three to four working animals are replaced by a tractor in real life. Our own investigation at the state level, leaving aside abnormal cases, has confirmed those findings. At the all-India level, we have come to the conclusion that a tractor replaces not more than 12 working animals. At this rate, if more than 70 million working animals (1992) were to be replaced by tractors, 6 million tractors would have to be in place to carry on agricultural production in India. Annually, it will cost about 20 million tonnes of diesel, the burning of which in running the tractors would cause over 6 million tonnes CO_2 emission, which gets prevented annually by using animal energy in Indian agriculture.

APPENDIX TABLE 6.1: DENSITY OF DRAUGHT ANIMAL POPULATION AND NUMBER OF TRACTORS IN SIX MAJOR STATES IN INDIA, 1972–92

S. No.	States	Items	1972	1982	1992
A	Low density states				
1	Punjab	(i) Population of draught animals (nos.)	1,426,000	1,435,000	787,000
		(ii) Number of tractors (nos.)	41,200	106,500	213,000
		(iii) Net sown area (000 hectare)	4,086	4,202	4,139
		(iv) Density* of draught animals	34.89	34.15	19.01
		(v) Density* of tractors	1	2.53	5.1
2	Haryana	(i) Population of draught animals (nos.)	940,000	841,351	736,400
		(ii) Number of tractors (nos.)	17,900	61,402	156,210
		(iii) Net sown area (000 hectare)	3,555	3,596	3,474
		(iv) Density* of draught animals	26.44	23.39	21.19
		(v) Density* of tractors	0.5	1.71	4.49
B	Medium density states				
1	Andhra Pradesh	(i) Population of draught animals (nos.)	5,386,000	4,859,802	5,644,261
		(ii) Number of tractors (nos.)	5,800	21,450	56,887
		(iii) Net sown area (000 hectare)	11,077	11,034	10,466
		(iv) Density* of draught animals	48.62	44.06	53.92
		(v) Density* of tractors	0.05	0.19	0.54
2	Tamil Nadu	(i) Population of draught animals (nos.)	4,317,000	3,664,522	2,751,193
		(ii) Number of tractors (nos.)	4,900	14,546	29,500
		(iii) Net sown area (000 hectare)	6,332	5,259	5,813
		(iv) Density* of draught animals	68.17	69.68	47.32
		(v) Density* of tractors	0.08	0.27	0.51

(contd.)

APPENDIX TABLE 6.1 (contd.)

S. No.	States	Items	1972	1982	1992
C	High density states				
1	Uttar Pradesh	(i) Population of draught animals (nos.)	13,874,000	14,930,000	12,392,000
		(ii) Number of tractors (nos.)	13,400	142,200	346,600
		(iii) Net sown area (000 hectare)	17,194	17,226	17,259
		(iv) Density* of draught animals	80.69	86.67	71.8
		(v) Density* of tractors	0.07	0.83	2.01
2	West Bengal	(i) Population of draught animals (nos.)	4,845,000	5,519,000	5,561,000
		(ii) Number of tractors (nos.)	-	2,000	6,500
		(iii) Net sown area (000 hectare)	5,565	5,341	5,334
		(iv) Density* of draught animals	87.06	103.33	104.25
		(v) Density* of tractors	-	0.04	0.12

Notes: Private tractors used in Agriculture. *Density per 100 hectare of net sown area.

Sources: Population and no. of tractors are taken frc n Mishra and Sharma (1990) for the year 1982 and population of working animals and tractors in 1972 and 1992 are taken from *Indian Livestock Census, 1977* and *1992*. Net sown area taken from CMIE, *Agriculture*, 2000.

7. Land Saving Due to the Livestock Production System

7.1 Introductory Remarks

As noted in Chapter 4, India's livestock production system saves on land in two ways: (*i*) by using agricultural by-products as animal feed and (*ii*) by producing dung output, a part of which is used by the society as domestic fuel. If the by-products were to be replaced by produced feed, feed-grains or green fodder, land will be required to produce these. Similarly, if dung output used as domestic fuel were to be replaced by firewood, land will be required to produce it. Either way, the land saving that occurs is beneficial, both environmentally as well as economically. Land being a natural resource, any activity that reduces or displaces its use is a nature conserving, environment protecting activity. Its economic value can be appreciated in the following way. India has reached her upper limit of the net cultivated area. Its human population in the year 2001 was a billion plus, and is growing. Any additional demand on that fixed land area from the livestock production system will have to be at the expense of food grains and fibre demand of India's growing population. The economic value of the land saved due to the livestock production system can be seen in terms of food grains output that will have to be foregone if that land were to be shifted to production of feed-grains or green fodder. The issue of valuation has been discussed in a subsequent chapter. For the present, our concern is how much of the land area is in fact saved. We begin with the estimation of the land area saved due to recycling of agricultural by-products as animal feed, followed by estimation of area saved from dung-cake output used as domestic fuel.

7.2 Estimate of Land Saving Due to Recycling of Agricultural By-products as Animal Feed

The procedure of estimation, at the risk of some repetition, is as follows (see also Chapter 4). The feed consumption per animal per day of by-products of different categories of bovines, concentrates and dry fodder, have been converted into their dry matter equivalents. These in turn have been converted into their energy values. The sum of the two gives us the gross energy intake of the animal in a particular category from the by-products. We then find out how much of cultivated green fodder on dry matter basis would be required to replace the gross energy intake of the animal from the by-products. From there, given the dry matter fraction of green fodder and the yield of green fodder per unit area, we have calculated the land area per animal per year in different categories that would be required to replace by-products feed by cultivated green fodder. Given the livestock populations in different categories, the next step is to apply the per animal rate to the respective populations to obtain the land that will be required by each category of population, and their sum total. The required parameter values and their sources are discussed in the following section.

7.2.1 PARAMETER VALUES

The per day per animal feed consumption of different types of feed is given in Jain and Singh (1990) (see Appendix Table 3.3). Their dry matter contents are given in Appendix Table 5.1. The energy values in calories, in other words, per day gross energy intake of the animals from by-products feed on dry matter basis are given in Table 7.1. The energy value in million calories per kg of green fodder on dry matter basis, and also of concentrates and dry fodder are the same as given in Table 5.3 of Chapter 5. The amount of green fodder on dry matter basis required to replace the gross energy intake from by-products (column 3) is given in column (4). The figures in column (4) are obtained by dividing the figures in column (3) by energy value per kg of green fodder (4.34 M cal.) on dry matter basis. The equivalent amounts of green fodder on green basis are given in column (5) of the table. The figures are obtained by applying green fodder to dry matter fraction ratio (1.0:0.25). For calculating the land area saved, the yield of green fodder per hectare had been taken to be 31.18 tonnes. It is the weighted average of yields under rainfed (25 tonnes) and irrigated conditions (50 tonnes) (Mishra and Sharma 1990). The weights are the irrigated and unirrigated area under fodder crops at all-India level in 1992–3.

TABLE 7.1: PARAMETER VALUES

Categories of population	Gross energy intake from by-products* per animal per day: Concentrates (M cal)**	Dry fodder (M cal)	Total (M cal)	Green fodder required to replace energy from by-products on dry matter basis (kg/day/ animal)	Required amount of green fodder on green matter basis (kg/day/ animal)
	(1)	(2)	(3)	(4)	(5)
Cattle					
Cow in-milk	1.07	16.64	17.71	4.08	16.32
Dry cow	0.32	14.55	14.87	3.42	13.68
Cattle adult male	1.09	18.19	19.28	4.44	17.76
Cattle young stock	0.17	7.19	7.36	1.69	6.76
Buffaloes					
Buffalo in-milk	3.73	21.28	25.01	5.76	23.04
Dry buffalo	0.92	17.34	18.26	4.2	16.8
Buffalo adult male	0.69	18.07	18.76	4.32	17.28
Buffalo young stock	0.14	6.77	6.91	1.59	6.36

Notes: *On dry matter basis. **M cal = million calories

Sources: For energy value of feeds on dry matter basis and dry matter fractions see Table 5.3 in Chapter 5; for per day per animal feed consumption data, see Appendix Table 3.3; for per day per animal feed consumption on dry matter basis, see Appendix Table 5.1.

7.2.2 THE RESULTS

The computational results based on the parameter values just discussed and the bovine population of 1992 are presented in Table 7.2. As can be seen from col. (3), if by-products feed were replaced by green fodder a total 1449 million tonnes of green fodder will be required annually, of which about 69 per cent will be required for feeding the cattle population, and 31 per cent for feeding the buffalo population. The total land area that would be required to produce this order of green fodder works out to 46.44 million hectare. This is the area that is annually saved by recycling agricultural by-products as animal feed. The country has been allocating around 5 to 6 per cent of net sown area for fodder crop production annually. In 1992–3 the area allocated was 8.41 million hectare out of the total net sown area of 142.51 million hectare. If we add to it the area required for producing green fodder in order to replace by-products feed, the

TABLE 7.2: LAND SAVING DUE TO AGRICULTURAL BY-PRODUCTS USED AS ANIMAL FEED, 1992

Categories of population	Population (million)	Per animal green fodder required to replace by-product feed (tonnes/year)	Total green fodder required col. (1) × col. (2) (MT)	Land area required to produce replacement amount of green fodder (m. ha)
	(1)	(2)	(3)	(4)
Cattle				
Cow in milk	31.562	5.95	187.79	6.02
Dry cow	32.799	4.99	163.66	5.24
Cattle adult male	74.489	6.48	482.68	15.48
Cattle young stock	65.734	2.46	161.70	5.18
Total cattle	204.584	-	995.83	31.92
Buffaloes				
Buffalo in-milk	25.892	8.40	217.49	6.97
Dry buffalo	17.914	6.13	109.81	3.52
Buffalo adult male	8.086	6.30	50.94	1.63
Buffalo young stock	32.314	2.32	74.96	2.40
Total buffalo	84.206	-	453.2	14.52
Total bovines	288.790	-	1449.03	46.44

Note: Yield of green fodder is 31.18 tonnes per hectare.

total required area would be about 55 million hectare, i.e. about 39 per cent of the net sown area. That is indicative of the order of importance of agricultural by-products as a source of animal feed in India.

7.3 Land Saving Due to Dung Used as Domestic Fuel

Methane emission from burning of dung-cake has been dealt with in Chapter 5. In the present section we deal with its positive environmental side. Besides being a sustainable biomass source of energy, production of dung-cake does not make a demand on real resources, except for some labour for making and drying of cakes and its storage. Production and supply of the alternative biomass fuel, namely fuel wood, on the other hand makes a demand on real resources, particularly land. The land saving that occurs due to dung-cake output used as fuel is the land area that would be required to produce and supply an equivalent quantity of fuel wood, if dung-cake output were to be replaced by fuel wood.

The values of the required parameters and their sources for estimating dung-cake output have already been given in Chapter 5 in the context of methane emission. However, for the convenience of the reader, the estimated quantity of wet dung output and the values of the parameters have been put together in Table 7.3. The dung evacuation rates for different categories of animals for estimating wet dung output are the same as given in Appendix: A Note on the Estimation of Dung Evacuation Rate. The sources of other parameters are given at the base of Table 7.3. Three parameters—thermal equivalence between dung-cake and fuel wood, yield of fuel wood, and the gestation lag in its production—however, need some observations.

The thermal equivalence is from a publication of the Khadi and Village Industries Commission (KVIC 1983, reprint 1999). The Commission has been engaged in promoting installation of biogas plants across the country. As part of that promotional activity the Commission has prepared a two-way table of replacement value of different fuels (see Appendix Table 7.1). According to this table 1.0 kg of fuel wood replaces 3.539 kg of dung-cake. It is rare to find a study on the economics of fuel wood plantation and harvesting. Based on a few projects in Gujarat, Patel (1985) has reported dry fuel wood yield of 100 tonnes per hectare under what he calls 'green revolution condition', and 30 tonnes per hectare under rainfed conditions. What he meant by 'green revolution condition' is intensive plantation culture with close planting of saplings,

TABLE 7.3: DUNG OUTPUT AND RELEVANT PARAMETER VALUES

S. No.	Items	Values
1	Total production of wet dung (million tonnes) in 1992–3.	682.19
2	The proportion of wet dung utilized for fuel (%).	40.00
3	Proportion of moisture in wet dung (%)	75.00
4	Replacement rate of fuel wood for dung-cake in terms of thermal energy	1:3.54
5	Yield of fuel wood in rainfed condition (tonnes per hectare)	30.00
6	Gestation lag between planting and harvesting of fuel wood trees (years)	3

Sources: Dung evacuation rate (see Appendix: A Note on the Estimation of Dung Evacuation Rate) and *Indian Livestock Census, 1992* for item no. 1; Central Statistical Organisation (CSO) Govt. of India for item no. 2; Garg et al. (1971) for item no. 3; Khadi and Village Industries Commission (1983) for item no. 4; Patel (1995) for item nos. 5 and 6.

irrigation, and application of high doses of chemical fertilizers. The gestation lag between planting and harvesting under both types of projects was 3 years. We have used yield under rainfed conditions, since this is what corresponds with normal practices in the country. Farmers generally grow fuel wood trees under rainfed conditions along fixed boundaries and lands not suitable for intensive crop cultivation. Though we have adopted 3 years' gestation lag, other conditions remaining the same, it may vary according to the species of fuel wood trees.

The results of the exercise are presented in Table 7.4. In column 1 are given dung-cake outputs attributable to the populations (1992) under respective categories of cattle and buffaloes. The corresponding replacement amounts of fuel wood are given in column 2. The figures under column 3 on land area required to produce the fuel wood are given just to show what difference gestation lag makes to the area required. The figures in column 4

TABLE 7.4: LAND SAVING DUE TO USE OF DUNG-CAKE AS DOMESTIC FUEL, 1992

Categories of population	Dung cake output (MT)	Fuel wood required to replace dung-cake (MT)	Land area required to produce replacement amount of fuel wood (million hectare)	
			With 1 year gestation lag	With 3 years gestation lag
	(1)	(2)	(3)	(4)
Cattle				
Cow in-milk	8.932	2.52	0.084	0.252
Dry cow	7.150	2.02	0.067	0.201
Cattle adult male	21.300	6.01	0.200	0.600
Cattle young stock	10.45	2.95	0.098	0.294
Total cattle	47.832	13.50	0.449	1.347
Buffaloes				
Buffalo in-milk	9.140	2.58	0.086	0.258
Dry buffalo	5.410	1.53	0.051	0.153
Buffalo adult male	1.680	0.47	0.016	0.048
Buffalo young stock	3.970	1.12	0.037	0.111
Total buffaloes	20.200	5.70	0.190	0.570
Total bovines	68.032	19.20	0.639	1.917

are the actual land area required to produce replacement amounts of fuel wood. At the aggregate level, dry dung-cake output worked out to over 68 million tonnes. If this were to be replaced by fuel wood the country will require to produce 19.2 million tonnes of fuel wood in addition to whatever amount it produces otherwise. Finally, to produce this amount of fuel wood, about 2 million hectares of land will have to be constantly kept under fuel wood plantations. This then is the land area saved due to dung-cake output used as domestic fuel.

7.4 Summary and Conclusions

To sum up, any activity that saves or displaces a natural resource is a nature conserving, environment protecting activity. India's livestock production system serves land in two ways: (*i*) by using agricultural by-products as animal feed, and (*ii*) by using a part of the dung output as domestic fuel. The social opportunity costs of both, the by-products and the dung-cake, are virtually zero. In this chapter we have made an attempt to estimate the land resource the country saves on account of both the activities. The land saving on account of by-products feed is considerably large. If the by-products feed were to be replaced by green fodder, as much as 1,449 million tonnes of the latter will be required annually. And, to produce it, 46.44 million hectare of land will be required annually. The land saving on account of dung-cake output of 68 million tonnes used as fuel is relatively small, about 2 million hectare to produce the replacement quantity of fuel wood of 19 million tonnes. The total land saving due to the livestock production system, thus, works out to over 48 million hectare. Its order of importance can be gauged by noting that it is as much as 39 per cent of the country's net sown area of 142 million hectare.

APPENDIX TABLE 7.1: REPLACEMENT VALUES OF DIFFERENT FUELS

Name of fuel	Unit	Gobar gas 1m^3	Kerosene 1 litre	Firewood 1 kg	Cow-dung-cakes 1 kg	Charcoal 1 kg	Soft coke 1 kg	Butane 1 kg	Furnace oil 1 litre	Coal gas 1m^3	Electricity 1 kWh
Gobar gas	m^3	1.00	1.61	0.29	0.08	0.69	0.62	2.31	2.40	0.85	0.21
Kerosene	litre	0.62	1.00	0.18	0.05	0.43	0.39	1.43	1.49	0.53	0.13
Firewood	kg	3.47	5.60	1.00	0.28	2.38	2.17	8.21	8.33	2.95	0.74
Cow-dung-cake	kg	12.30	19.83	3.54	1.00	8.44	7.64	28.39	29.48	10.44	2.62
Charcoal	kg	1.46	2.35	0.42	0.12	1.00	0.91	3.37	3.50	1.24	0.31
Soft coke	kg	1.61	2.59	0.46	0.13	1.01	1.00	3.71	3.85	1.36	0.34
Butane*	kg	0.43	0.70	0.13	0.04	0.30	0.27	1.00	1.04	0.37	0.09
Furnace oil	kg	0.42	0.67	0.12	0.03	0.29	0.26	0.96	1.00	0.35	0.09
Coal gas	kg	1.18	1.90	0.34	0.10	0.81	0.73	2.79	2.82	1.00	0.25
Electricity	kg	4.70	7.58	1.35	0.38	3.22	2.93	10.85	11.26	3.99	1.00

Notes: For equivalents read vertical columns 1 m^3 (cubic meter) = 35.315 cubic feet.

* Butane: Bottled domestic cooking gas sold as Indane, Bharat (Burshane), H.P. (Esso), etc.

Source: Khadi and Village Industries Commission (1983).

8. Saving of Chemical Fertilizers Due to Dung Used as Manure

8.1 INTRODUCTION

The sub-model presented in Section 4.3 of Chapter 4 is meant to determine the amount of carbon dioxide emission prevented due to part of the dung output used as organic manure in place of chemical fertilizers. The sub-model, at the same time, determines the amounts of various chemical fertilizers that would be required to replace dung manure for the same amounts of plant nutrients the dung manure has. In the present chapter we are concerned with the estimation of chemical fertilizers that would be required in order to replace the plant nutrients made available from part of the dung output used as manure. It is worthwhile noting that dung as organic fertilizer, apart from providing plant nutrients, is considered superior to chemical fertilizers from the environmental angle in the sense that unlike chemical fertilizers it does not adversely affect the natural fertility of the soil, and protects the soil born micro-organisms. Section 2.2 presents the required data for estimation purposes and their sources. The results are provided in Section 8.3. The last section concludes with some remarks.

8.2 THE DATA

The output of wet dung from cattle and buffalo populations (1992) at the evacuation rates given in Appendix: A Note on the Estimation of Dung Evacuation Rate has already been estimated at 682 million tonnes (Chapter 5, Section 5.2). The proportion of dung output used as manure (60 per cent), the moisture content of wet dung (75 per cent), and their sources too have been discussed earlier. The nutrient contents per unit of chemical fertilizers are very large in comparison to nutrients per unit of dung manure (see Table 8.1). Therefore, the replacement amounts of fertilizers are expected to be small.

TABLE 8.1: NUTRIENT CONTENTS OF MANURE AND CHEMICAL FERTILIZERS

S. No.	Items	Values
1.	Proportion of wet dung utilized for manure (%)	60.00
2.	Proportion of moisture in wet dung (%)	75.00
3.	Fraction of plant nutrients in dung manure:	
	(1) Nitrogen	0.005
	(2) Phosphorous	0.002
	(3) Potash	0.005
4.	Fraction of plant nutrients in chemical fertilizers:	
	(1) Nitrogen in ammonium sulphate	0.206
	(2) Phosphorus in super phosphate	0.444
	(3) Potash in muriate of potash	0.660

Sources: Central Statistical Organisation (CSO) Govt. of India for item nos. 1 and 2; Garg et al. (1971) for item no. 3; Goel et al. (1973) for item no. 4.

8.3 The Results

The result as presented in Table 8.2 shows that out of the total wet dung output in 1992, about 103 million tonnes on dry matter basis was used as manure. If it were to be replaced by chemical fertilizers,

TABLE 8.2: CHEMICAL FERTILIZERS SAVING DUE TO DUNG USED AS MANURE, 1992

(*in million tonnes*)

Categories of population	Dung manure output on dry matter basis	Ammonium sulphate required to replace nitrogen in dung manure	Super phosphate required to replace phosphorous in dung manure	Muriate of potash required to replace potash nutrient in dung manure
	(1)	(2)	(3)	(4)
Cattle				
Cow in milk	13.41	0.326	0.060	0.102
Dry cow	10.66	0.259	0.048	0.081
Cow adult male	32.10	0.780	0.144	0.244
Cow young stock	15.58	0.379	0.070	0.118
Total cattle	72.31	1.744	0.322	0.545
Buffaloes				
Buffalo in-milk	13.69	0.333	0.062	0.104
Dry buffalo	08.12	0.197	0.037	0.062
Buffalo adult male	02.54	0.062	0.011	0.019
Buffalo young stock	05.91	0.144	0.027	0.045
Total buffaloes	30.26	0.736	0.137	0.230
Total bovines	102.57	02.480	0.459	0.775

the replacement amounts of the latter work out to about 2.5 million tonnes of ammonium sulphate, 0.46 million tonnes of super phosphate, and 0.78 million tonnes of muriate of potash. The distributions according to animal categories are incidental. These are meant to keep a check on the consistency of the results with the results in the preceding chapters involving the wet dung output of the bovine population. The replacement amounts of the chemical fertilizers are apparently small. But their monetary values could be substantial, an aspect that is discussed in the following chapter.

8.4 Concluding Remarks

The saving of chemical fertilizers due to dung used as manure, though important, does not give its full environmental value, the way saving of land due to dung-cake used as fuel does. In the latter case saving is of a natural resource. In the former case it is saving of manufactured chemical compounds using mineral resources. It is indeed the saving of mineral resources required to manufacture the substitute amounts of chemical fertilizers which ought to count for the true environmental value of dung manure. Mineral resources are exhaustible. To the extent they are used, they get depleted. Dung used as manure ultimately prevents their depletion to the extent it displaces substitute chemical fertilizers based on the minerals.

9. Valuation of the Environmental Effects

9.1 The Approach to Valuation

During the last two decades a vast body of literature has grown relating to various aspects of environment, environment related development policies, and strategies for protection of the environment. This literature is put under the broad rubric of 'environmental economics'. Environmental economics, *inter alia*, is concerned with the valuation of the environmental effects of economic activities, policies, and strategies in the *ex ante* as well as in the *ex post* situations. A closer examination of the literature shows that the approach to valuation of environmental effects is essentially that of the conventional cost–benefit analysis. The cost–benefit analysis technique itself took shape in the post Second World War period, when the state played a major role in economic development, especially in the developing and former socialist countries. During that period which ended with the disintegration of the former Soviet Union in 1991, macro and inter-sectoral planning models were used to determine investment allocations. Cost–benefit analysis was used to evaluate public investment projects in order to enable macro level planners or policy makers to choose among competing projects, worthwhile projects which could fill up the pre-determined investment of a sector or sub-sector of the economy. In those heady days of development planning, few people bothered about the environmental consequences of a development project. One of us who did a cost–benefit evaluation of a marine fisheries project and another of an intensive cattle development project (ICDP) hardly cared about the projects' environmental effects (Mishra and Beyer 1976; Mishra 1978). However, that as it may, the cost–benefit analysis as a technique of valuation has proved its usefulness and versatility. Be it a project, a policy, a strategy, or a production system, all of these are brought into the cost–benefit accounting framework for valuing and assessing their environmental impacts. Indeed, it has been used to estimate the global costs and benefits arising out of possible climatic change 50 to 100 years hence due to global warming that may be caused by the accumulation of greenhouse gases in the earth's atmosphere (for a survey, see Cooper 2000). The technique's versatility also points to its potential for misuse. It is not difficult for one to underestimate cost and overestimate benefit, or vice versa, especially in *ex ante* situations where most entities involved in the calculus have to be projected from the present into the future. This is not the place, however, to make a review of the cost–benefit analysis technique. Our purpose is limited to its valuation aspect only. The following few paragraphs are meant to provide a brief sketch of the method of valuation.

9.1.1 VALUATION METHOD

Consider a dairy project. Like any other project, the dairy project can be viewed as a technical relation between inputs and outputs, a process or activity of transforming inputs into outputs. Now, suppose that apart from human labour, crop straw as animal feed is input to the project. And, there are two project outputs: milk and methane.

Our problem now is to value the straw input and the outputs. A farmer who participates in the project may have to buy a part of his straw requirement from other farmers at the prevailing market price of straw in or around his village. The value of this input for this farmer, in other words, his cost, is equal to the quantity of straw purchased multiplied by the market price of straw. Consider now the value of straw input to the project from the angle of all farmers collectively, that is, from the angle of the society as a whole. Straw is a by-product of agricultural production. Its production does not require additional allocation of real resources, such as land, labour, and capital. Therefore, its cost of production is nil. Hence its value from the social angle is zero. What if straw has alternative uses for the society? For instance, it may be required as a packaging material in the economy. In that case its social value can not be zero. What is its social value then? Its value is equal to its *opportunity cost*. The opportunity cost arises when an input has been withdrawn from one use to another. In the present case when straw has an alternative use as a packaging material, withdrawing it for the dairy project at the margin would mean a sacrifice of its value as a packaging material, which will cause a fall in the output of the packaging industry to that extent. That sacrifice of value, in other words, the *value foregone* in the packaging industries due to the withdrawal of the straw to the dairy project is its social opportunity cost or its value. The general proposition is that if there is an input, such as straw, whose real resource cost of production is nil but it has an alternative use in the economy, it has a positive value called its opportunity cost, which is equal to the value foregone in its alternative use.

There is another social angle from which the value of straw needs to be looked at. Instead of looking at its value from the dairy project in question, consider it as an input to the country's dairy production system. As we know from the outset its resource cost of production is nil. But, what if it was to be replaced, for one or another reason, by feed-grains or cultivated green fodder? Obviously, the production of the latter types of feed will require allocation of real resources, in particular land. For a given area of land available for cultivation of crops, that would mean shifting of land required for production of these types of feed from its present alternative uses, for instance, area allocated to food grains production. Clearly, the value of straw presently used as feed is as much as the value of food grains output that will have to be foregone if land were shifted to production of the replacement amount of feed-grains or green fodder. This is the sort of land saving that otherwise occurs due to agricultural by-products used as feed, a point that we have frequently come across in the preceding chapters.

Let us now get back to the dairy project for valuation of the outputs. The valuation of milk output is straightforward. If there is a milk market in which competitive conditions prevail, the market price of milk is its unit value. Milk output multiplied by the price is the value of output. Farmers participating in the project may retain a part of the output for their own family consumption. This portion too can be given an imputed value at the same market price. If there is no market price available, a survey inquiry on *willingness to pay* among potential consumers can be conducted to find the maximum price consumers are willing to pay. The project's milk output can then be evaluated at this price.

The other project output, namely methane, does not accrue to anybody in our society. Therefore, both from the private as well as social angle, its value is zero. However, methane output escapes to the earth's atmosphere and becomes a part of the global output of methane. The global emission of methane along with emission of other greenhouse gases at the current rates, if unchecked, is predicted to change the earth's climate about a century from now, making the temperate zones warm and tropical zones warmer still. Countries in the temperate zones are likely to be the gainers and those in the tropics are likely to be the losers due to the climate change (Cooper 2000). India's agriculture, for instance, may suffer; food production may fall. Superficially one might say that the value of the expected fall in the food output is the value (cost) of greenhouse gases emission from India. But that is wrong, since the expected fall will be due to global emission rather than emission exclusively from India.

In the environmental economics literature it is quite common to take examples of pollutants or affluents from industrial projects for illustrating the negative values (costs) of such output. If there is technology available which can prevent the output of a pollutant or convert it into harmless products, the cost of that technology is supposed to be the negative value of the pollutants' output.

As regards our dairy project, the output of methane cannot be prevented, since it is part of the metabolism of the ruminants. What is, however, possible is to reduce the methane output per unit of milk output. As we have noted earlier a feed-grain based feeding technology produces much less CH_4 emission per unit of milk output than one based on roughages, green or dry. Can we then say that the cost of feed-grains based technology is equal to the negative value of the amount of CH_4 emission reduced on the analogy of the pollutants' case? The answer is: no. For, in the pollutants case the technology does not make any adverse environmental impact—at least, it is supposed so. Not so with the feed-grains, whose production, apart from other things, requires land, a natural resource. Thus, there is an environmental trade-off. On the one hand you reduce methane output, on the other you increase the demand on land. The problem of valuation of methane output, therefore, remains unresolved.

In Chapter 5, we have demonstrated that the buffalo based milk production system is superior to the cow based system. For the same, or almost identical, combination of feed inputs the buffalo based system produces more milk per unit of gross energy intake and less methane per unit of milk produced compared to the cow based system. Consider now buffaloes and cows as two separate technologies of milk production. Think of replacing cows by buffaloes as the buffalo technology reduces CH_4 output per unit of milk produced. One might then say that the value (negative) of the reduced amount of CH_4 output is equal to the investment cost of the required number of buffaloes, ignoring for the moment the transition period. Note that the buffaloes, or for that matter cows, can not be manufactured with no loss of time, or in the same one year period. But replacement of cows will have another adverse environmental impact: displacement of the working bullocks' stock by the required number of tractors and other agricultural machinery run on fossil fuel. Manufacture of the machinery will make a demand on non-renewable mineral resources. In addition, there will be greenhouse gases emission, CO_2 in particular, from fossil fuel required to run the machines and to manufacture them. Thus, there is an environmental trade-off here too. It would not be right, therefore, to say that the value (cost) of the reduced amount of CH_4 emission due to buffalo technology is equal to the investment cost of buffaloes. In view of these complexities we have refrained from making any attempt to find out the negative value (cost) of the methane output from India's livestock production system. The values of its positive environmental effects are presented in the following section.

9.2 Value of the Positive Environmental Effects

The estimate of values of the positive environmental effects of the livestock production system are presented in Table 9.1.

Column (1) figures give the savings under various items in physical quantities. These have been pooled together from the preceding chapters. For estimating the value of land saving we have taken rice as the representative foodgrain produced across the country. Rice yield per hectare is the all-India average. It is clear from Table 9.1 that the most important value contributions of the system come through land saving due to recycling of agricultural by-products as animal feed. The sum of Rs 76,357 crore should be interpreted as the value that will have to be *foregone* if the energy input to the system from by-products feed were to be replaced by energy from cultivated green fodder. Next in importance comes the non-renewable fossil fuel energy, diesel saving by animal energy use in Indian agriculture. At the current retail price in Delhi, the saving is as much as Rs 35,357 crore, a little less than half of the value on account of by-products feed, leaving aside the CO_2 emission prevented due to this. The third in importance is the lowly 'cow-dung' cake, the poor man's home energy. Its environmental value works out to Rs 3144 crore. In other words, if dung cake were to be

TABLE 9.1: VALUES OF POSITIVE ENVIRONMENTAL EFFECTS

Items	Area (million hectare)	Foodgrains* output (rice) (MT)	Value (crore Rs.)
Land saving			
Due to by-Products feed	46.56	87.6259	76,357.21
Due to dung used as fuel	1.917	3.6077	3143.75
Total	48.477	91.2336	79,500.96
Saving of fertilizers due to dung manure	Quantity (MT)	Price (Rs per tonne)	
Ammonium sulphate	2.48	4000	992.00
Super phosphate	0.459	8300	380.97
Muriate of potash	0.775	3700	286.75
Total	3.714	16,000	1659.72
Diesel saved due to animal energy use	19.50	18,132	35,357.00
Total value	-	-	116,517.68

Note: *Assumed yield of foodgrains (rice) is 1.88 tonnes per hectare.

Sources: For domestic price for rice at Delhi wholesale market, see Agricultural Prices in India, Ministry of Agriculture, Govt. of India, 1996–7; for yield of rice in 1996–7, see CMIE (1999); for fertilizer prices, see USDA, Fertilizer Marketing News, September 1999; for diesel price, the Delhi retail prices (current) are taken.

replaced by fuel wood produced and supplied by fuel wood plantations, this much of value will have to be foregone by the society. The last in the line of importance comes the saving of chemical fertilizers due to dung used as manure. At the current prices of fertilizers the value of the replaced amounts of fertilizers works out to about Rs 1660 crore.

The total value of the positive environmental effects of India's livestock production system is as high as Rs 1,16,518 crore. Compare this with the total value of output of the livestock sector as a whole. In 1997–8, at 1993–4 prices it amounted to Rs 78,682 crore (GOI 1999). Even if the former was deflated, that is brought back to price levels of 1993–4, it will be far larger than the value of the output of the system. Few of us could ever imagined that the value of the positive environmental contribution of the country's livestock production system could be so large until it has been demonstrated. We hope that we have demonstrated it. Note, however, that the orders of magnitude involved rather than the precise numbers given in the table are important for appreciating the system's environmental value.

9.3 Methane Balance of the Livestock Production System

Among the people concerned with greenhouse gases emission, global warming, and climate change, the ruminant livestock populations around the world are marked as a source of methane emission. India's livestock being the world's largest in number, no one can ignore India for methane emission from her livestock sources. Table 9.2 presents the methane balance of India's livestock production system. As mentioned in Section 9.1.1, we have refrained from attempting to find the negative environmental value of the methane output of the system.

The figures in Table 9.2 are again pooled together from the preceding chapters. The entry in the last but one row of the table is questionable, since we have as yet not been able to find the reliable data required for estimating CO_2 emission that occurs in the manufacturing processes of chemical fertilizers.

It is clear from Table 9.2 that the predominant source of methane emission is enteric fermentation of feed, accounting for 96 per cent of the actual emission from the system. The emission prevented by the system in rows (4) and (5) are the methane equivalents of the CO_2 emission in terms of global warming potential. The emission of CH_4 prevented due to animal energy use in agriculture (0.30 million tonnes) is only a little less than the total from 'manure management' and burning of dung-cake (0.412 million tonnes). We expect that the

TABLE 9.2: METHANE EMISSION BALANCE OF THE LIVESTOCK PRODUCTION SYSTEM, 1992

S. No.	Source of emission	Quantity (MT)
1	Enteric fermentation	9.355
2	Manure management	0.202
3	Burning of dung-cake	0.210
	Total emission (1+2+3)	9.767
4	Emission prevented due to animal energy in place of fossil fuel	(-) 0.300*
	Balance	9.467

Note: * This is methane equivalent to 6.19 million tonnes of CO_2 prevented, methane having 21 times higher global warming potential (GWP) than CO_2 (see section 4.5 Chapter 4 and Table 6.5).

Source: Tables 5.3, 5.5 and 5.7.

balance of CH_4 emission, once the figure for emission prevented due to dung manure in place of chemical fertilizers is put in the table, will be around 9 million tonnes. For the interested reader the inventory of emissions from the livestock production system in CO_2 equivalents is given in Appendix Table 9.1.

9.4 Concluding Remarks

To sum up, the approach to valuation of environment effects of economic activities is none other than the standard cost–benefit analysis technique. However, complexity of the environmental linkages is such that even if one is able to quantify them, it is not always easy to work out their values, be it cost (negative effect) or benefit (positive effect). For instance, methane output from the livestock sources of a country is a global output. Its effects are global which can not be distributed according to the countries of origin. Similarly, a technology that may reduce CH_4 emission from enteric fermentation turns out to have other adverse environmental effects.

Nevertheless, the value of the positive environmental effects of India's livestock production system is truly large, worth as much as Rs 1,16,518 crore, much more than the value of output of the livestock sector (Rs 78,682 crore in 1997–8). Major environmental values are those of the agricultural by-products used as animal feed (Rs 76,357 crore), and animal energy used in agriculture (Rs 35,357 crore). Due to the difficulties pointed out earlier we have refrained from valuing methane output from the system. Of the total actual emission in 1992 (9.767 million tonnes) as much as 96 per cent is from enteric fermentation of feed (9.355 million tonnes). The balance of methane emission taking credit for emission prevented due to animal energy use works out to 9.467 million tonnes. Once we are able to take credit for emission prevented due to dung used as manure in place of chemical fertilizers, we expect the methane balance of the country's livestock production system to be around 9 million tonnes.

APPENDIX TABLE 9.1: GREENHOUSE GASES INVENTORY OF THE LIVESTOCK PRODUCTION SYSTEM, 1992

(*in MT*)

S. No.	Sources	CH_4 emission	CO_2 emission	CO_2 equivalent
1	Enteric fermentation	9.3555	-	196.46
2	Manure management	0.202	-	4.242
3	Burning of dung cake	0.21	-	4.41
4	Fuel oil saved due to animal energy	-	-6.19	6.19
5	Emission prevention due to dung manure	-	- ?	
	Total CO_2			198.922

Note: The methane has been converted into carbon dioxide @ 21GWP (1 kg of methane = 21 kg of carbon dioxide in terms of global warming potential). The total of CO_2 equivalent does not take credit for CO_2 emission prevented due to dung used as manure.

10. Comparison with a Developed Country, the United States of America

10.1 The Approach

The environmental effects of a livestock production system occur both on the input as well as output side of the system. An effect can be beneficial or harmful to the environment. An output can have both sorts of effects. The output of dung (manure), for instance, depending upon the way it is 'managed' causes more or less methane emission, a harmful effect, since methane contributes to global warming. At the same time, if output of manure is used as organic fertilizer its effect is beneficial, since it substitutes for chemical fertilizers. An input like feed-grain has a similar dual effect. A high level of feed-grains fed to animals, as in a feedlot system, reduces the incidence of methane emission from enteric fermentation, which is a beneficial effect. But the production of feed-grains using fossil-fuel energy causes emission of CO_2 and other greenhouse gases. Also, Fossil fuel is a non-renewable, unsustainable source of energy. Furthermore, the production of feed-grains makes a demand on arable land, which is a limited natural resource.

The environmental effects of a livestock production system, as noted in Chapter 2, fall into two categories: *global* and *local* (regional/national), depending upon the space or area of their impact. Greenhouse gases' emission *due to the system* obviously belongs to the global category. Due to the system means inclusive of the emission the system causes indirectly, for instance CO_2 emission from fossil fuel use in the production of feed-grains, or CH_4 emission from burning of dung-cake as biomass fuel. Incidentally, in preparing the national inventory of greenhouse gases according to the IPCC *Guidelines* (1995), CH_4 emission from the output side alone, that is from enteric fermentation and manure management, is debited to the livestock sector. Indirect emissions from the input side such as CO_2 emission from fossil fuel use in feed production is debited to the energy sector. While this creates no problem as far as national inventory is concerned, it mars the intersectoral comparison of greenhouse gases emission.

While the global environmental effect of all livestock production systems is the same (greenhouse gases emission causing global warming), that is not the case with regard to local environmental effects. Local effects may differ sharply depending upon the social, cultural, and economic settings in which production systems are organized. This can be noticed in the contrasting local environmental effects manure output and its disposal has in the production settings of a developed and developing country. In most developing country situations where livestock production, integrated with agriculture, is organized in small sized family farms, manure output is small. This output is used on the farms as organic fertilizer or as biomass fuel, or both as in India. Its local environmental effect in either case is beneficial.

In contrast, in most developed countries, livestock production is separated from agriculture. It is organized on a large scale in farms and factories (beef cattle feedlots, pigs and poultry production facilities). Animals are fed large quantities of quality feed, supplemented with minerals and other additives. Accordingly, large volumes of dung, called 'mountains of manure' by the neighbourhood people, are produced. By and large it is a waste product of the system, which poses a disposal problem. It is disposed or 'managed' in two ways: (*i*) stored in covered lagoons and slurry-pits, and (*ii*) spread on fields. No matter which way it is disposed, its local environmental effect is one of polluting the ambient air and the underground and surface water. Noxious gases like ammonia and hydrogen sulphide released from the manure permeates the air, causing health problems. In the Netherlands, ammonia emission from manure is as much as 87 per cent of the total annual emission. Ammonia concentration in the atmosphere beyond a point causes acid rains (de Haan et al. 1997). There is also nitrate pollution of underground water through leaching of nitrogen and surface water through run-off when manure is spread on the fields. The undigested excreted part of the feed has so much of nutrients (nitrogen, phosphoric acid) that plant life is unable to absorb it. Manure has most of the heavy metals (copper, zinc, nickel, manganese) that are added to animal feed as growth supplements. Animals are able to digest only 5 to 15 per cent of these additives, and the balance gets deposited on the fields. It contaminates crops and thus poses risk to human health (de Haan et al. 1997). Its accumulation has the potential of turning the soil toxic and infertile. Run-off from the fields flushes the metals into the surface water sources, where it has been found to interfere with the reproduction of aquatic plants and animals. In parts of the developed world, such extensive damages to the local environment have reached crisis proportions. In the state of Missouri in the US, for instance, citizens have set up a Missouri Rural Crisis Centre, which monitors damages and campaigns against 'factory farms', and in favour of 'family farms' (for a general account, see Cantrell et al. 2001).

International comparison of the environmental effects of livestock production systems has so far been largely confined to the global effects. There again, it has been narrowed down to a comparison of methane emission from the output side of the systems, that is from enteric fermentation and 'manure management'. The World Resources Institute (WRI), for example, makes country-wise annual estimates of the latter, and ranks the countries in order according to the total amounts of emission. India expectedly figures at the top of their list since India's livestock holding is the largest in the world.

Such a gross comparison may serve the political purposes of some of the developed countries, which have been putting pressure to bring large developing countries like India into binding international protocol on the reduction of greenhouse gases, though they among themselves have failed so far to enter into one. But it is patently unfair since it does not take into account the differences in the human populations of the countries under comparison although methane emission from livestock sources is supposed to be anthropogenic. For a fair comparison, therefore, country-wise methane emission from livestock sources need to be normalized to a common anthropic base, that is reduced to per capita level. In fact, in general wherever measurement or estimate of a characteristic pertains to different bases, it needs to be normalized to a common base, in other words reduced to a common denominator, before it can validly be compared. From this angle, the above sort of comparison suffers from methodological error too.

Be that as it may, a systems approach requires that whenever any two livestock production systems are to be compared for their environmental effects, not only global but also local effects, both from the output and input side, should be taken into account. Prior to that it is necessary to ensure that the production systems are comparable. Two systems are comparable if they produce the same socially useful product such as milk using the same species of animals or close relatives (bovines). Differences in factors of production (land, labour, capital) deployed, feed and feeding levels, etc., do not matter. For comparison purpose these differences can be taken care of by normalizing the factors of production and operating inputs to a common output base, say per tonne of milk produced. The differences in the animal-breeds (highly selected Holstein, Jersey breeds *vs.* common Indian cow) are, however, difficult to account for. A system of milk production based on highly selected breeds of cattle, other things remaining the same, retains its inherent comparative advantage. So is the case with climatic differences.

We stated above that any two livestock production systems are comparable provided they produce the same socially useful product, and use the same species of animals. But in the real world things are different. Consider cattle based production systems, for instance. In the developed countries, milk and beef production are separate specialized activities. Accordingly, at any point of time there are two separate stocks of cattle: 'milk cattle', and 'beef cattle'. While animals culled from milk cattle stock go to slaughter, their role in beef production is very marginal. In most developing countries, there is no such specialization. The same cattle (buffalo) stock is used for production of milk as well as beef. Some of the calves, heifers, dry cows, and adult males are culled out of the stock and slaughtered for beef. In such vastly different situations how can one compare a cattle based production system in a developing country with the one in a developed country? The problem, however, is not insurmountable. A comparison can still be attempted by making use of the following steps. The first step is to aggregate the milk and beef production systems of the developed country into one system. Specifically, it means aggregation of the resources, inputs, and outputs of the two sub-systems. The aggregate system, as if, produces two products, milk and beef, which is what is also produced by the cattle based production system in the developing countries. This step meets the first condition of comparability, that is the two systems in question produce the same socially useful product(s). Since milk and beef outputs of the two systems are likely to be in different proportions, and, in any case, they cannot be added together, the second step is to convert milk and beef outputs of either system into equivalent amounts of calories or proteins, and add them up to get the total amount of calories or proteins either system produces. Let there be equivalent outputs of proteins on either side. We are now ready to take the final step. The final step is to normalize the systemwide resources, inputs, etc., to a common base of protein output. Once normalized this way the two systems become comparable. Fortunately, this problem is not faced in India, as beef production from cattle (buffalo) is a residual, marginal activity here. Cattle (buffalo) are raised primarily for milk production, although cows are also supposed to replace the stock of working bullocks.

Just as to make any two livestock production systems comparable their resources, inputs, etc., are to be normalized to a common output base, so too their environmental effects need to be normalized to the same common output base. An underlying assumption here is that all the environmental effects of a system, global as well as local, are quantifiable. However, quantitative estimates of some of the effects, especially of the local ones, may not be readily available. In such situations only a general qualitative comparison can be attempted.

10.2 The Scope of Comparison

Following the approach described above we make an attempt in this chapter to compare the milk production systems of India and the United States of America, and their respective environmental effects. We have chosen the United States for the simple reason that, besides being a large sized country, the US is fairly representative of the conditions of livestock production in the developed world. The milk production systems of the two countries are, for the most part, comparable if we ignore: (*i*) beef production from India's bovine stock, and (*ii*) the bullock replacement function of cows. Apart from these, there is another factor that will have to be ignored, but for a different reason. More than 90 per cent of the US milk cows belong to the highly selected Holstein-Friezian breed. Their hereditary milk production trait is vastly superior to that of Indian cows or buffaloes. This inherent difference between the two milk production systems due to the difference in the genetic make-up of the animals gives the American system an inherent advantage, since this difference will remain unaccounted for when we normalize the two systems to a common milk output base.

In addition to these two provisos, there is a third proviso which needs to be noted. As regards resources and inputs to the systems, we have brought into the picture only animal feed and fodder, and land resources deployed in their production and for grazing. Labour and capital resources and other inputs have been left out of consideration. The underlying assumption is that animal feed (fodder) is the most important and critical input to livestock production processes, and it is the basis of almost all direct or indirect environmental effects on the output as well as input side of a livestock production system.

Keeping the above provisos in mind, a comparison of the milk production systems of India and the US is made in Section 10.4. A broad account of the US livestock production system as a whole is given in Section 10.3, primarily for the benefit of those Indian readers who may be unfamiliar with it. The environmental effects of the two systems are presented and discussed in Section 10.5. Section 10.6 is devoted to a summary statement of the main findings of the study.

10.3 An Outline of the US Livestock Production System

In 1992, the US had a livestock population of over 174.4 million heads. Cattle and hogs (pigs) together accounted for about 90 per cent of the total livestock. Cattle, numbering about 98 million heads, had the largest share, at 56 per cent of the total. The number of hogs was over 58.2 million, their share being 33.4 per cent of all livestock. In the remaining 10 per cent of the livestock, in order of importance, sheep and lamb stood at the top, numbering about 10.8 million, followed by horses (5.8 million), and goats (2 million). The US has large poultry bird population also, with their number being more than 1642 million in 1992 (See Appendix Table 10.1).

It is interesting to note that the cattle population in the US is divided into two streams: (*i*) dairy cattle and (*ii*) beef cattle. In 1992, the share of beef cattle in the total cattle population was as much as about 82 per cent, their number being more than 79.5 million. In the total dairy cattle population of about 18 million heads, milk cows numbered a little over 9.7 million heads, the remaining being the replacement stock.

The US livestock population data for 1996, the latest available to us, when compared with the data for 1992 shows that there has been a change in favour of beef cattle. Their number increased to about 103.5 million in 1996 from the 1992 level of 79.5 million. During the same period, dairy cattle stock declined marginally from about 18 million to 17.6 million. The population of milk cows declined from 9,728 thousand in 1992 to 9416 thousand in 1996. Interestingly, the population of hogs, a major source of red meat, also declined from 58 million in 1992 to 56 million in 1996. On the other hand, there was a substantial increase in the number of poultry birds from 1642 million in 1992 to 2097 million in 1996 (see Appendix Table 10.1). These changes should not be taken for a set trend, but rather as indicative of a short to medium term change in the preference pattern of the people for livestock food products.

The US livestock system is geared to the production of milk, various types of meat, and poultry products. In line with the two separate streams of cattle mentioned above, milk and beef production are specialized activities. The stock of 'milk cows' a highly selected one, is mostly comprised of Holstein–Friezian breed. Whereas its female calves (heifers) born during the year are available for replacement of the stock, male calves (steers) are available for beef production. But their number is comparatively too small to make a significant addition to the beef cattle population. 'Beef cows' too seem to be a selected lot, whose 'calf crop' has the trait of faster bodymass accumulation, specially when at a certain age it is 'shipped' to feed lots, where it is subjected to intensive grain feeding before slaughter. The ratio of the *flow* of cattle and calves slaughtered to the beef cattle *stock* is in the range of 40 per cent or more. In 1992, for a total beef cattle stock of about 80 million heads, the number slaughtering about 34.5 million heads, out of which 1.4 million were calves. A relatively small number of calves at a tender age are slaughtered for veal. Like the beef feedlot system, pig and poultry production are also organized on factory lines. Sows being prolific breeders, it is not surprising that the number of hogs slaughtered during a year is far larger than the hog population at the beginning or end of the year—in other words, the *flow* to *stock* ratio is more than one. In 1992, for a hog population of 58 million, the number slaughtered was 95 million heads. The same is the case with poultry production, with a *flow* to *stock* ratio of 4:1 in 1992.

In 1992, the US produced 68.6 million tonnes of milk, 18.6 million tonnes of red meat, and 16 million tonnes of poultry meat. Red meat is estimated on 'carcass equivalent' basis, i.e. weight of the animal minus entrails, head, hide, and internal organs. Poultry production is on liveweight basis. The composition of the red meat output is as follows: Beef, about 10.5 million tonnes; Pork, 7.8 million tonnes; Veal, 0.14 million tonnes; Lamb and Mutton, 0.16 million tonnes. Clearly, beef and pork account for 98 per cent of the red meat, their respective shares being 56 and 42 per cent of the total output. It may be noted that in the US system, poultry production is almost as important as red meat production.

For producing the milk and meat outputs at the scales noted above, the US deploys vast amounts of feed and forage, and the land acreage required to produce them. Animal feed as usual is classified into two parts: concentrates and roughages. Concentrates consist of two parts: feed-grains and meals. Whereas feed-grains are directly produced from land, meals are by-products. Roughages too consist of two parts: 'harvested roughages' such as corn (maize) silage and hay made from clover, beet, alfalfa, etc., and forages animals graze on in pastures.

According to an estimate of the Economic Research Service (ERS) of the US Department of Agriculture (USDA), the total feed consumed by the livestock and poultry in 1992 from all the three sources (concentrates, harvested roughages, and pastures) amounted to over 447 million tonnes in 'equivalent feeding value of corn'. Corn value probably refers to nutrient value of other feeds and forages *vis-à-vis* corn feed. Nutrient value of a feed or forage is measured in terms of its 'total digestible nutrients' (TDN) content. Thus, if x amount of a feed has y amount of TDN, and if x amount of corn has the same y amount of TDN, then 'corn value' of x amount of the feed in question is equal to x. In short, using an equivalence scale between corn and other feeds in terms of their TDN, all feeds can be converted in corn equivalents, and summed up to get 'equivalent feeding value' of all the feeds consumed in terms of corn.

The composition of the total feed consumed in corn equivalents (447.25 million tonnes) was as follows: concentrates 211.38 million tonnes, harvested roughages 74.39 million tonnes, and forages from pastures 161.48 million tonnes. In the concentrates, by-product meals are included. By-product meals amounted to 36.62 million tonnes in 1992. Total concentrates minusby-product meals gives us the feed-grains consumed equal to 174.76 million tonnes in equivalent corn. Note that hereafter feed refers to its 'equivalent feeding value of corn'.

How much land acreage was allocated to produce this order of feed-grains, harvested roughages, and forages? The area under the latter two categories is easily obtained from land utilization statistics of the US for 1992. As for the area allocated for feed-grains, some explanation of the manner we have arrived at it is necessary. Feed-grains include a set of four grains: corn, sorghum, barley, and oat. Not all of the annual output of these grains goes as animal feed. We have the statistics on output disposal to various uses of these grains, one of which is what is termed as 'disappearance as feed and residuals'. The quantity of any of these grains under this head should be closest, though not identical, to the quantity consumed as feed. For 1992, we have the quantities of all the four grains disposed under this head, and their shares of the total. Applying these shares to the total amount of feed-grains actually consumed, 174.76 million tonnes in 1992, gives its distribution among the four feed-grains. Dividing the respective quantities by the yield per hectare of the corresponding grains, we have obtained the land area under each grain, the sum of which gives us the total land area accountable to the total feed-grains consumed in 1992. It works out to 24.19 million hectare, almost 90 per cent of which was devoted to corn. By adding to this the area under harvested roughages (24.37 million ha.) and pasture land (208 million ha) we get the total area devoted to production of feed and forage consumed in 1992, which is about 257 million ha. (see Appendix Table 10.2). To conclude, a US livestock population of 174.4 million heads and the poultry birds numbering 1642 million together consumed over 447 million tonnes of feed, about half of which was concentrate in the form of feed-grains and by-product meals in 1992. To obtain this order of feed the US devoted as much as 257 million ha of land. This is a far cry from the Indian situation indeed.

What is the feed consumed per day per animal in the US, say by a milk cow, a heifer, or a poultry bird? The key to the answer lies in the *animal-equivalence scale in terms of their feed consumption.* On dry matter basis, feed consumed by an average milk cow during a year is supposed to correspond to one standard 'animal unit'(see Appendix Table 10.3). Now, suppose that an average hog's consumption of the same feed on dry matter basis is one-quarter of that of the average milk cow, then the hog is equal to 0.25 animal unit. In this way average animal of different categories of livestock are converted into their equivalent animal unit(s), or 'factors', so to say. Multiplying the livestock number in a given category with its 'factor' gives its equivalent number of animal units. By summing these number of animal units over all types of livestock and poultry one gets the total number of animal units during the year in question.

The ERS of the USDA has developed equivalence scales in animal unit(s) corresponding to four different types of feed fed to livestock and poultry. These are: (*i*) grain; (*ii*) roughages; (*iii*) grain and roughage; and (*iv*) high protein (meal supplements). For a given type of feed, the livestock population in a given group when multiplied by its scale factor yields its equivalent number of animal units. For example, consider the scale for feed-grain. Suppose x is the scale factor for hog, and hog population in the year in question is X, then its equivalent number of grain consuming animal units (GCAU) is x X. In the same way one can compute GCAUs corresponding to all types of livestock and poultry. Similarly, applying the scale for roughage feed one can compute roughage consuming animal units (RCAU) for all types of livestock and poultry. Now, the total of a feed consumed by the livestock and poultry during a year can be distributed among different groups in the proportions of their GCAUs or RCAUs, as the case may be. It is then easy to calculate feed consumed per animal unit, and per head of the respective groups.

For 1992, for which estimates of total feed consumed by the livestock and poultry populations are available, we carried out an exercise to find out feed consumed per animal unit, and per head of the different groups of the population. Feed-grains and by-product meals in corn value have been allocated in proportions of GCAUs, and roughages in proportion of RCAUs. We have also allocated land area under feed-grains and roughages in the same respective proportions in order to find out area per head of different groups of livestock and poultry. The results are presented in Appendix Tables 10.4a, b, c, and d. Feed consumed per head per day from various sources in corn values for selected groups of livestock and poultry, and land deployed per head per year are given in Appendix Table 10.5.

Appendix Table 10.5 shows that an average dairy cow in the US is fed more than 20 kg of feed in corn value per day, more than 11 kg from feed-grains, and meals and the rest from roughages. 'Beef cattle on feed', that is those in the feedlot system, are fed as much as

14 kg of grain, and about 3 kg of meal per day. Even the average pig receives about 3 kg of feed a day in corn value, over 2 kg in the form of grain, and the rest from concentrate meal and roughages in that order.

Land per animal deployed for feed production is equally large, especially for dairy and beef cattle. Remember that land for feed falls under three categories: (*i*) land for feed-grains, (*ii*) harvested roughages, and (*iii*) pasture land. For purposes of allocation among roughage consuming animal units we have treated the latter two as a single category, since the equivalence scale for roughages is inclusive of harvested roughages and forages from pastures. Land per head under the two categories of feed-grains and roughages for different groups of livestock and poultry is given in the Appendix Table 10.4. It will be observed from the table that total land per dairy cow works out to 3.62 ha, out of which about 0.5 ha is for feed-grains production. The corresponding figure for 'beef cattle on feed' is much lower at 1.16 hectare per animal, but a larger part of it, about 0.7 ha, is for feed-grains production. Typically the young stock of cattle in the US—replacement heifers, steers, and calves above a certain age—are kept on pasture grazing and provided minor supplements of feed-grains and meals. Accordingly, the land area per head of this group is in the range of 2 to about 3 hectare, most of it under roughages.

To conclude, the US with its relatively small livestock population produces fairly large amounts of milk and meat. The high level of productivity of her livestock population is based upon two factors: (*i*) highly selected breeds of animals (poultry) for purposes of milk and meat (beef) production, and (*ii*) high level of nutritive feeding of the animals. As noted earlier according to the ERS of the USDA, a livestock population of 174.4 million heads together with poultry birds numbering 1642 million consumed as much as 447 million tonnes of feed in corn equivalent in 1992, about half of which was concentrate, mostly feed-grains. In the same year, an average milk cow consumed more than 20 kg of feed a day in corn equivalent, about 56 per cent of which were from concentrates. Similarly, 'a beef cattle on feed' was fed about 18 kg of feed a day, as much as 14 kg in the form of feed-grains. It may be noted that corn accounts for 90 per cent of all feed-grains consumed, and between 60 to 70 per cent of the year's corn production in the US is utilized as animal feed. To facilitate such order of feed consumption and feeding levels, the US deploys a vast area of land for feed production. In 1992, in addition to 208 million hectares of pastures for grazing, over 24 million hectares were put under corn, forage grasses, clover, etc. for making silage and hay (harvested roughages). And, feed-grains consumed during the year, 175 million tonnes (in corn equivalent), according to our estimate, required another 24 million hectares of land, bringing the total to 256 million hectares of land.

10.4 Comparison of Milk Production Systems: India and the United States

Against the above general background of the US livestock production system, we now focus attention on a comparison of India's milk production system with that of the United States. Although milk production is the economic goal of the systems, two other outputs inevitably flow from the systems as joint products: methane and 'manure' (dung). These, as we have seen, have direct bearing on the environment. These are treated in the following section, where the systems are compared for their environmental effects. Here the comparison is confined to the economic aspects only.

The relevant statistics for comparison are presented in Tables 10.1 and 10.2. Figures in Table 10.1 relate to dairy population, milk output, milk yield, feed intake per animal, and related parameters such as per animal energy intake, TDN intake, and land input. Note particularly that this way the two systems are normalized on per animal basis. Following the approach described in Section 10.1, we have at the same time normalized the systems on per unit milk output basis. The results are presented in Table 10.2.

It is clear from Tables 10.1 and 10.2 that comparison of feed inputs poses difficulty, except possibly in the case of concentrates, if one ignores its composition on either side. This is because the US estimates of feeds consumed are available but in their 'equivalent feeding value of corn'. One way to overcome the difficulty, and this is what we have done, is to convert equivalent corn values in their TDN contents and make them comparable with TDN per animal and per unit milk output on the Indian side. Chemical analysis of Indian corn grain (maize) reveals that its TDN content is 84.9 per cent (Sen et al. 1978). Applying this percentage to corn equivalent of feeds consumed we obtain its TDN content. It is likely that American corn has higher nutrient contents. To that extent, figures given in

Tables 10.1 and 10.2 would be underestimate. For the Indian side, the TDN figures are based on Jain and Singh (1990). There is a second difficulty which, however, remains. TDN, as we know, consists of two parts: digestible crude protein (DCP) and digestible energy (DE). Depending upon the type of feed from which TDN is derived, the proportions of DCP in TDN vary. Concentrates have a substantially higher proportion of DCP than roughages. Total digestible nutrients from crop residues, especially of cereal crops, consists almost wholly of DE. This is why a ration with higher proportion of concentrates is considered better for animal nutrition and milk production.

As regards land input required for producing feed-grains, and roughages (harvested + grazed) consumed by an average milk cow in the US, we have estimated it in the manner described in the preceding section. From there it is easy to work out land input per unit milk output (see Table 10.2), given the milk cow population and total milk output. In the case of India, under the assumption that all of the cultivated green fodder output is consumed by the bovine stock only, we have allocated the area under green fodder in proportion to the green fodder consumed by different categories of cattle and buffalo populations (see Appendix Table 10.7). This way once we have the land areas corresponding to dairy cows and buffaloes (in-milk/dry) it is easy to calculate land input per dairy animal or per unit milk output. It needs no saying that land area corresponding to dry fodder (crop residues) is zero. The area under pastures in India in comparison to the United States is miniscule, 11 million ha *vs.* 208 million ha in 1992. Incidentally, while in India pasture lands are either under the public or village community ownership, in the US pastures are under the private ownership of the farmers themselves. Now, even if we unrealistically suppose that all pastures in India are available for grazing by the bovine population alone (288.79 million heads in 1992), one can easily see that the pasture area per dairy animal (in-milk/dry) or per unit milk output would be extremely negligible.

With the above explanatory note, let us now comment on the figures given in Table 10.1. It is obvious that as compared to India, the US produces an incredibly large amount of milk with far fewer milk cows.

In 1992, with just over 9.7 million milk cows the US produced over 68 million tonnes of milk as against a little over 55 million

TABLE 10.1: MILK PRODUCTION SYSTEMS: INDIA AND THE US

S. No.	Particulars	India	United States
I	Dairy cows/buffaloes (in million)		
(*a*)	Cows		
(*i*)	Cow in-milk	31.562	9.728
(*ii*)	Dry cows	26.231	-
(*iii*)	Total	57.793	9.728
(*b*)	Buffaloes		
(*i*)	Buffalo in-milk	25.892	-
(*ii*)	Dry buffaloes	14.383	-
(*iii*)	Total	40.275	-
II	Milk Production (million tonnes)		
(*i*)	Cow milk	24.338	68.554
(*ii*)	Buffalo milk	31.044	-
(*iii*)	Total	55.382	68.554
III	Milk yield/animal/day (kg)		
(*i*)	Cow in-milk	2.11	19.40
(*ii*)	Milch cow	1.15	-
(*iii*)	Buffalo in-milk	3.28	-
(*iv*)	Milch buffalo	2.11	-
IV	Feed intake/animal/day (kg)		
(*a*)	Cow in-milk		
1	Concentrates	0.273	11.23
(*i*)	Feed-grains	-	9.29*
(*ii*)	By-product meals	-	1.94*
2	Roughages	-	8.79*
(*i*)	Crop residues (dry fodder)	5.01	-
(*ii*)	Green fodder	5.09	-
(*b*)	Buffalo in-milk		
1	Concentrates	0.947	-
(*i*)	Feed-grains	-	-
(*ii*)	By-product meals	-	-
2	Roughages		
(*i*)	Crop residues (dry fodder)	6.41	-
(*ii*)	Green fodder	8.772	-
V	Total digestible nutrients (TDN) intake/ animal/day (kg)		
(*i*)	Cow in-milk	3.49	16.99
(*ii*)	Buffalo in-milk	5.24	-
VI	Gross energy intake (GEI) animal/day (million joule)		
(*i*)	Cow in-milk	97.26	331.93(299.5**)
(*ii*)	Buffalo in-milk	144.57	

(contd.)

VII	Land for feed/animal/year (ha)		
1	Cow in-milk		
(*i*)	Feed-grains	-	0.46
(*ii*)	Crop residues (dry fodder)	-	-
(*iii*)	Green fodder	0.032	3.16+
(*iv*)	Pasture	0.034	-
2	Buffalo in-milk		
(*i*)	Feed grains	-	-
(*ii*)	Crop residues (dry fodder)	-	-
(*iii*)	Green fodder	0.056	-
(*iv*)	Pasture	0.058	-

Notes: * Feed intake per animal/per day for US are expressed in equivalent feeding value of corn.

(1) US has no buffaloes in its livestock production system.

(2) Pasture land found negligible for cow in-milk and buffalo in-milk (per animal per year).

\+ This includes land for roughages and pasture both.

** IPCC estimate of Gross Energy Intake of US dairy cow (MJ/per animal/day).

Sources: (1) Indian Livestock Census, 1992.

(2) Agriculture Statistics, USDA, 1994.

(3) Basic Animal Husbandry Statistics, 1999.

(4) Statistical Abstract of the United States, 1995.

(5) Jain and Singh, 1990.

(6) Agriculture Statistics, Feed Consumed, USDA, 1994.

(7) Sen, Ray and Ranjhan, 1978.

(8) Feed Outlook, USDA, 1995.

(9) Agriculture Statistics, Grain and Feed, USDA, 1994.

tonnes of milk with about 32 million cows and 26 million buffaloes in-milk. Accordingly, the milk yield of a US cow (19.4 kg per day) is about 9 times that of an Indian cow in-milk (2.11 kg per day), and about 6 times that of a buffalo in-milk (3.28 kg per day). But then, ignoring the fact that the US dairy cattle is a highly selected elite stock, the quantity and quality of feed fed to the US cows is also incomparably larger and better. Although we do not have the information on the quantity of roughages fed to an average US cow, its corn equivalent is about 9 kg a day. Adding to that 11 kg of concentrates, the total works out to 20 kg a day in corn equivalent, about 56 per cent of which is in the form of concentrates. Quite in contrast, the total feed intake of an average Indian cow in-milk, on raw material basis, is about 10.4 kg a day, and that of a buffalo in-milk 16 kg a day. More important, about 97 per cent of the feed in the case of cows in-milk and 94 per cent in the case of buffaloes in-milk is in the form of roughages. Furthermore, crop residues (dry fodder) make up 50 per cent of the roughages in the case of cows, and 42 per cent in the case of buffaloes in-milk. Thus, in contrast to American cows, India's cows and buffaloes make milk almost wholly out of roughages, between 40 to 50 per cent of which consists of crop residues.

Consistent with the contrasting patterns of feed intake, an average American cow's intake of TDN is 17 kg a day while that of an Indian cow in-milk is about 3.5 kg and of a buffalo in-milk 5 kg a day. Expectedly the latter is predominantly comprised of DE, drawn as it is from roughages. In fact the DCP portion in these TDN values is no more than 7 per cent. On a rough guess we believe that the DCP content in the TDN intake of an American milk cow would not be less than 20 to 25 per cent.

Finally, as pointed out in the preceding section, about half a hectare (0.45 ha) of land is deployed to produce the feed-grains, mostly corn consumed per year by an average milk cow in the US. And, the area under roughages, harvested roughages and pastures, works out to 3.16 ha per milk cow, most of it being pasture area. The area per cow in-milk (0.034 ha) or per buffalo in-milk (0.056 ha) in India for producing the green fodder consumed is apparently negligible in comparison, and so is the area under pastures, as noted earlier.

It is clear from the preceding discussion that an average US milk cow produces many times more milk than an Indian cow or buffalo in-milk, consumes many times more feed (nutrients), and requires many more times land area for producing the feed it consumes. When systems are normalized on per animal basis, it is precisely such differences that are revealed. But these differences need to be accounted for, in other words eliminated for revealing the comparative efficiency of the systems in input–output terms. The results of such an exercise are presented in Table 10.2. Prior to commenting on the figures it is necessary to note that feed intake (nutrients) of dry animals are included in the total feed consumed by the populations on either side. In the preceding discussion we did not pay attention to this aspect because dry animals do not produce milk in any case, and the systems' comparative efficiency was not under consideration. On the US side we do not have definitive statistics on the proportion of dry cows in the total (9.7 million in

TABLE 10.2: INPUT PER UNIT OF MILK OUTPUT IN INDIA AND THE US, 1992

S. No.	Input items	India	United States
I.	Feed (kg/tonne of milk output)		
1.	Cow based system		
(*i*)	Concentrates	159.42	583.00*
(*a*)	Feed-grains	-	482.00*
(*b*)	By-product meals	-	101.00*
(*ii*)	Roughages	-	
(*a*)	Crop residues (dry fodder)	4094.00	456.00*
(*b*)	Green fodder (including silage and hay)	3973.00	-
(*c*)	Farm grazing	-	-
2.	Buffalo based system		
(*i*)	Concentrates	327.3	-
(*a*)	Feed-grains	-	-
(*b*)	By-product meals	-	-
(*ii*)	Roughages	-	-
(*a*)	Crop residues (dry fodder)	2835.00	-
(*b*)	Green fodder (including silage and hay)	3775.60	-
(*c*)	Farm grazing	-	-
3.	Bovine based system		
(*i*)	Concentrates	253.5	-
(*a*)	Feed-grains	-	-
(*b*)	By-product meals	-	-
(*ii*)	Roughages	-	-
(*a*)	Crop residue (dry fodder)	3388.8	-
(*b*)	Green fodder (including silage and hay)	3862.26	-
(*c*)	Farm grazing	-	-
II.	Total digestible nutrients (TDN) (kg/tonne of milk output)		
1.	Cow based system	2711+	854.00
2.	Buffalo based system	2353.4+	-
3.	Bovine based system	2510.7	-
III.	Land (ha/tonne of milk output)		
1.	Cow based system		
(*i*)	Concentrates	-	-
(*a*)	Feed-grains	-	0.067
(*b*)	By-product meal	-	-
(*ii*)	Roughages	-	0.449
(*a*)	Crop residues	-	-
(*b*)	Green fodder	0.072	-
(*c*)	Pasture land	-	-
	Total	0.072	0.516
2.	Buffalo based system		
(*i*)	Concentrates	-	-

(contd.)

(*b*)	Feed-grains	-	-
(*c*)	By-product meals	-	-
(*ii*)	Roughages	-	-
(*a*)	Crop residues	-	-
(*b*)	Green fodder	0.069	-
(*c*)	Pasture land	-	-
3.	Bovine based system		
(*i*)	Concentrates	-	-
(*a*)	Feed-grains	-	-
(*b*)	By-product meals	-	-
(*ii*)	Roughages	-	-
(*a*)	Crop residues	-	-
(*b*)	Green fodder	0.070	-
(*c*)	Pasture land	-	-

Note: * Quantity expressed in equivalent feeding value of corn.
+ Estimates for breeding cows, breeding buffaloes, and bovines.
Source: Table 10.1.

1992), but the generally quoted figure is about 10 per cent. In comparison, India has an overload of dry animals—45 per cent of the breeding cows as well as buffaloes according to the 1992 Livestock Census. That happens because of poor nutrition, nutrients being predominantly drawn from roughages.

The figures in Table 10.2 show that for producing one metric tonne of milk the US system uses 583 kg of concentrates (in equivalent feeding value of corn) as against 159.42 kg by India's cow based system, 327.3 kg by the buffalo based system, and 253.5 kg by the total bovine based system. In roughages, as noted earlier, comparison is not possible since we have only corn equivalent of roughages per tonne of milk produced (456 kg) in the US. There are two indirect evidences which, however, indicate that the amount of roughages could be pretty large. One is the land area deployed for roughages as we will see below. The other is the fact that an average milk cow in the US produces as much as 41 kg of 'manure' (dung) per day (refer to the following section). On the Indian side the amount of roughages per tonne of milk produced is quite large: 8 tonnes for milk from the cow based system, 7 tonnes for milk from the buffalo based system, and about 8 tonnes for milk from the bovine based system as a whole, crop residues accounting for about 50 per cent of the total. These figures along with the concentrates per tonne of milk translate into equally large amounts of TDN per tonne of milk produced: over 2.7 tonnes, 2.35 tonnes, and 2.51 tonnes per tonne of milk produced respectively from the cow, buffalo, and bovine based systems. On the other hand, if we ignore the difficulties in translating the corn equivalents of feed inputs into their TDN values, the US system uses remarkably less TDN per tonne of milk produced, just 854 kg, corresponding to total corn equivalent of all feeds amounting to a little over one tonne (1.039 tonnes) per tonne of milk produced. But the natural resource cost of this achievement is also remarkably high. The US system uses more than half a hectare of land (0.516 ha) for producing a tonne of milk compared to 7–8 cent of a hectare by the Indian system(s), that is about seven times more.

What do we conclude from the above, rather detailed description? When the Indian and the US system of milk production are compared on per animal basis sharp contrasts between the two come out on the surface. An average US milk cow produces about six to nine times more milk per day (19.4 kg) than an average Indian buffalo in-milk (3.28 kg) and cow in-milk (2.11 kg) in that order. The average US milk cow consumes more feed per day even when feeds are reduced to 'equivalent feeding value of corn' (20.02 kg) compared to the Indian buffalo (16 kg) and cow in-milk (10.4 kg), where feeds are on raw material basis, and more than 94 per cent of which are roughages. In the TDN consumed, the contrast turns out to be much sharper. The TDN consumed per day by a US milk cow (17 kg) is more than three times the TDN consumed by a buffalo (5.24 kg) and about five times the TDN consumed by an Indian cow in-milk (3.49 kg). Still more sharper differences appear when we consider land area deployed for producing the feed consumed annually by an average milk animal on either side. Per milk cow the US system deploys a total of 3.62 ha, of which about half hectare is for producing feed-grains, mainly corn consumed by the animal during a year. This is phenomenally large compared to the area deployed per Indian buffalo (0.058 ha) or per cow in-milk (0.034 ha), that is about 65 times as large as for a buffalo and 113 times as large as for a cow in-milk (see Appendix Table 10.6).

The above differences on per animal basis between the two systems do not tell us anything about the comparative efficiency of the systems. The comparative efficiency appears when we normalize the systems on per unit of milk produced. There, as regards feed inputs per unit of milk output the American system scores over the Indian system. However, as regards the land input per unit of milk output the Indian system scores over the American system.

Specifically, the American system uses 1.039 tonnes of feed in equivalent amount of corn per tonne of milk produced as against 8.226 tonnes, 6.937 tonnes, and 7.503 tonnes per tonne of milk produced respectively by the cow, buffalo, and the bovine based systems in India. The only caveat to add here is that the latter figures are on raw material basis and over 94 per cent of these consist of roughages. The same holds good when these figures are translated into their TDN values, although the differences get narrowed down substantially: 84 kg TDN per tonne of milk in the US system as against 2.3 to 2.7 tonnes TDN per tonne of milk in the Indian systems. As regards land input per tonne of milk produced, the Indian systems score is far higher, between 7 to 8 per cent of a hectare as against over 3 hectare. Indeed, this is the major environmental strength of the Indian livestock production system as we have seen in Chapter 7.

10.5 Environmental Effects of the Milk Production Systems

This section deals with the environmental effects of the milk production systems of India and the US in a comparative perspective. In Section 10.1 we had made a distinction between the *global* and *local* (regional/national) environmental effects of livestock production systems. Accordingly, we deal first with global effects followed by local effects. Recall that the global effect is in the form of greenhouse gases emission which causes global warming, and possible climate change. And, the local effects are those that have a positive or negative impact on the local environment.

10.5.1 Global Effect: Greenhouse Gases Emission

As in Section 10.4, for purposes of comparison we have normalized the greenhouse gases emission from the US and Indian milk production systems on per animal basis, and also on the basis of per unit milk output. The results are presented in Tables 10.3 and 10.4. Note that the figures on methane emission are inclusive of emission from 'manure management'. The figures for the US are based on the methane emission estimates made by the Environmental Protection Agency (US EPA 1990–8). The figures on the Indian side are our own estimates (see Chapter 5). According to the EPA, in 1992 total CH_4 emission from milk cows was 1.747 million tonnes, out of which 0.597 million tonnes was from 'manure management'. On the Indian side, total CH_4 emission from bovine milk production system was 4.22 million tonnes, out of which 0.091 million tonnes was from manure management. Specifically, and inclusive of emission from 'manure management', in 1992 the emission from the cow milk production system was 2.086 million tonnes, and from the buffalo milk production system 2.129 million tonnes. The US uses a considerable amount of fossil fuel energy in the production of feed-grains/harvested roughages, in irrigation, in making of silage, house warming for animals, etc. From the cost of feed-grains production data (USDA 1991) we have been able to obtain fossil fuel used in mechanical operations alone connected with feed-grains production per hectare of land (see Appendix Table 10.8). Using this information in conjunction with our estimates of land under feed-grains and feed-grains consumed by the population of milk cows in the US in 1992 we have worked out the total fossil fuel used, and its corresponding CO_2 emission. We have assumed the same rate of fossil fuel used in the production of harvested roughages. The latter has been reduced to CO_2 emission (and equivalent CH_4 emission) per animal and per unit of milk produced. The figures given in Tables 10.3 and 10.4 are underestimates, since

TABLE 10.3: CH_4 EMISSION PER ANIMAL PER YEAR

S. No.	On output side	Per animal/year (tonnes)	
	Methane emission (enteric fermentation + manure management)	India	United States
A	Cow based system		
(i)	Milk cow	-	0.180
(ii)	Cow in-milk	0.039	-
(iii)	Milch cow	0.036	-
B	Buffalo based system		
(i)	Buffalo in-milk	0.058	-
(ii)	Milch buffalo	0.053	-
C	Bovine based system		
(i)	Bovine in-milk	0.048	-
(ii)	Milch bovine	0.043	-
D	On input side		
(i)	CO_2 emission in production of feed-grains consumed	-	0.032
(ii)	CH_4 equivalent	-	0.002

Sources: Table 5.8 for India; *Agriculture Statistics*, USDA, 1997; USEPA (1990–8); *1991 Farm cost and return survey for corn*, USDA; IPCC (1995) for CH_4 equivalent.

these do not take into account the fossil fuel used by the US milk production system as a whole. As regards the Indian system, due to tractorization of agriculture there may be some fossil fuel used in the production of green fodder consumed by the dairy cows/ buffaloes, and corresponding CO_2 emission. We have, however, assumed it to be negligible because, as we have seen in Chapter 6, Indian agricultural operations are mainly based on animal energy use.

Looking at the figures in Table 10.3, it is obvious that CH_4 emission per animal per year from the US milk production system is pretty high, 180 kg, as compared to the Indian System—36 kg from the cow based system, 53 kg from the buffalo based system, and 43 kg from the overall bovine based system, that is between four to five times higher compared to Indian systems. But when we turn to CH_4 emission per unit of milk produced, there is a reversal of the positions, as shown in Table 10.4. The US system produces only 26 kg of methane per tonne of milk produced as against the Indian figures of 98 kg per tonne of milk by the cow based system, 73 kg by the buffalo based system, and 83 kg by the bovine based system. Thus, the Indian system produce three to four times more methane per tonne of milk produced than the US system. One way to compare the environment related efficiency of the two systems is to apply a set of two criteria: (*i*) milk produced per unit of TDN consumed,

TABLE: 10.4: CH_4 EMISSION PER UNIT OF MILK OUTPUT

S. No.	On output side	Per tonne milk output	
	Methane Emission (Enteric fermentation + manure management)	India	United States
1	Cow milk	0.098	0.026
2	Buffalo milk	0.073	-
3	Bovine milk	0.083	-
	On input side		
1	CO_2 emission in production of feed-grains consumed	-	0.005
2	Methane equivalent	-	0.0002

Sources: Emission data from USEPA, for the US; Emission data our own estimate for India; *Agriculture Statistics, 1997*, NASS, USDA for US milk output; *Basic Animal Husbandry Statistics, 1999* for Indian milk production; *1991 Farm cost and return survey for corn*, USDA for CO_2 emission; IPCC, (1995) for CH_4 equivalent.

and (*ii*) CH_4 emission per unit of milk produced, as we did in Chapter 5 while comparing the efficiency of the cow and buffalo based milk production systems. However, as we have seen in Section 10.4 there are difficulties in translating the feed intake of a US cow into its TDN value as different feeds are available only in their 'equivalent feeding value of corn'. Secondly, there is a certain degree of in-comparability between the TDN consumed on either side arising out of vastly different proportions of concentrates in the respective diets, hence in the DCP contents of their TDN equivalents (see Section 10.4). And, DCP is the critical factor in milk production.

Therefore, instead of TDN consumed we have chosen land input to milk production, a natural resource on which to base the environment related efficiency of the US and Indian systems. The general efficiency criteria now reads as follows: If a system A produces more milk per unit of land input than system B, and at the same time produces less CH_4 emission per unit of milk output than B, then A is more efficient than B. The estimates of these two parameters for the US cow based system and the Indian bovine based system are given in Table 10.5. It is clear that the Indian system produces about seven times more milk per hectare of land input (14.24 tonnes) than the US system (1.94 tonnes). At the same time it produces about three and a half times more methane emission per tonne of milk output (83 kg) than the U.S system (26 kg). The comparative position remains unchanged even when we exclude pastures and take into account only the land for feed-grains and harvested roughages on the US side. Milk produced per hectare of land goes up to 7.87 tonnes but still remains far too lower than that of the Indian system (14.24 tonnes). Methane emission per tonne of milk output on either side obviously remains as it is.

TABLE 10.5: TRADE-OFF BETWEEN LAND AND METHANE EMISSION

S. No.	Particulars	India	United States
1	Milk produced per hectare of land (milch bovines/milk cows)	14.24	1.94
2	Methane emission per unit of milk produced	0.083	0.026

Sources: *Basic Animal Husbandry Statistics, 1997* for land and milk production (India); *1997 Census of Agriculture and Agricultural Statistics, 1994*, USDA for land and milk production (US); *Inventory of US Greenhouse Gas Emission and Sinks: 1990–98* for methane emission (US); *Methane emission for India*, our own estimates.

Thus, in terms of the criteria stated above, neither system dominates over the other. Therefore, from the environmental angle one cannot say that the US milk production system is more efficient than the Indian system, or vice versa. There is a clear trade-off between the land deployed for milk production and CH_4 emission from production. In the US case, more land is deployed for quality feeding and there is less CH_4 emission per tonne of milk produced. Similarly, in the Indian case where there is less land deployed and greater dependence upon inferior quality feeding, there is more CH_4 emission per tonne of milk produced. The choice, *ceteris paribus*, is of course determined by the relative land endowment of the two countries whereas the US with its vast land resources can afford to produce less milk per unit of land and less CH_4 emission per unit of milk output, India cannot do so. It has been suggested that for developing countries like India with large population of cow and buffaloes, better quality feeding is a 'win–win strategy' (de Haan et al. 1997), since it permits more milk output with reduced number of animals, and CH_4 emission too is reduced in the process. But, as we have demonstrated, there is nothing like a win–win strategy. Reduced CH_4 emission has its land resource cost. A developed country like the Netherlands may claim that with this sort of strategy it produces more milk per unit of land and less CH_4 emission per unit of milk produced. But that claim is spurious. The Netherlands imports more than 60 per cent of its concentrate feed requirement from other countries, in particular cassava from a developing country like Thailand (de Haan et al. 1997). The major part of her land resource requirement of a high rate of milk production and low rate of CH_4 emission, therefore, is met by other countries.

10.5.2 LOCAL ENVIRONMENTAL EFFECTS

The local environmental effects of a livestock or milk production system are those that impinge positively or negatively on the local natural resources such as land, soil, water, and forest. Other effects include pollution of the local atmosphere and damage to local ecosystems. It is necessary to note that the extent and intensity of the local environmental effects largely depend upon the mode of production and production organization, that is, whether it is pastoral, small scale mixed-farming type or large scale, commercial production system. We are presently concerned with the local environmental effects of the latter two types, mixed farming system as practised in India, and large scale commercial farming system as practised in the United States. Whatever be the mode of production, just as in the case of global effects, local effects too may be on the input as well as output side of the system.

Manure (dung) is an outstanding output of the milk production systems in India and the US, and for that matter anywhere else too, which has local environmental effects. On the input side, animal feed, its form, and manner of procurement has an impact on the local land resources. Let us consider manure first. On the Indian side, we have seen in Chapters 7 and 8 how dung, whether used as organic fertilizer or biomass fuel has a positive impact on the local environment. As organic fertilizer, it replaces equivalent amounts of chemical fertilizers, maintains the natural fertility of the soil, and protects soil-born micro-organisms. As biomass fuel, it indirectly protects forests, trees, and bushes, and directly replaces equivalent amount of fuel-wood and saves land that would otherwise be required to produce the fuel-wood. We have a quantitative idea of these local effects of the dung output of India's milk production system. In 1992, dung output from dairy animals (milch bovines) used as manure on dry matter basis worked out to about 46 million tonnes. The total amount of different chemical fertilizers it replaced sums up to about 1.7 million tonnes (see Chapter 8, Table 8.2). The dry dung-cake output from the same animals used as biomass fuel was 31 million tonnes in 1992. And the land area it saved worked out to about 900 thousand hectare.

On the US side, we have precious little information on the utilization pattern of dung output from the milk production system, except that large scale concentrated production facilities produce large volumes of dung daily, and store it in lagoons or slurry-pits. An average US milk cow is supposed to produce 41 kg of wet manure a day (Gibbs et al. 1989). At this rate with 9.7 million milk cows the output of wet manure from the US milk production system works out to about 145.5 million tonnes in 1992. It is not clear whether this includes droppings during grazing on pastures. Whether it is included or not, the amount of droppings on the pastures ought to have similar local effects as the dung manure in the Indian case, namely replacement of equivalent amount of chemical fertilizers, maintenance of natural soil fertility, and protection of soil-born micro-organisms. That part of the dung output which is stored in lagoons and slurry-pits is known to have deleterious local effects, as was alluded to in Section 10.1. As for the United States as a whole we have no information on the intensity of such effects.

On the input side, animal feed to milk production systems invariably has a direct or indirect impact on the land resources of the country concerned. In Section 10.4, it was shown how the US uses an incomparably larger area of land to produce a tonne of milk (0.516 ha) than India does (0.070 ha). This remains true even when one excludes pastures lands accountable to the US milk cows (0.13 ha per tonne *vs.* 0.07 ha per ton). The plea could be that the density of milk cows being low, the natural status of pastures and remains unaffected, that there exists a symbiotic relationship between the milk cow population and the pastures, as happens to be the case in pastoral systems in which stocking density is lower than the carrying capacity of the pasture.

Underlying the above pattern of land use is the difference between the quality and quantity of feed fed to animals. Whereas the US uses high quality feed (feed-grains, silage, hay, and forages), to produce which it deploys land, India relies more on agricultural by-products feed requiring no land, supplemented by cultivated green fodder for which it deploys little land. This sort of economic choice, as stated earlier, is determined, *inter alia,* by the relative endowment of land resources of the two countries. From the purely environmental angle, that is from the angle of impact the two milk production systems make on the local environment, the Indian system is better as for producing the same amount of milk output it uses much less land. The less human exploitation of a natural resource for producing the same amount of a good or service there is, the better it is. Incidentally, it may be of interest to know that by recycling agricultural by-products to the milk production system India saved about 22 million ha of land (see Chapter 7, Table 7.2), while the US deployed a total of 35 million ha (including pastures) of land for its milk production systems in 1992 (see Appendix Table 10.6).

10.6 Summing Up

In the prevailing worldwide concern about greenhouse gases emission it is quite common to compare methane emission from livestock sources of one country with that of another. We have argued that such gross comparisons are not only unfair, but also methodologically incorrect, and limited in scope. In a proper systems approach it is necessary to take into account all environmental effects of a livestock production system, both on the input and output side of the system, and whether their impact is global (methane emission) or local (dung used as manure). Secondly, it is necessary to ensure that the systems are comparable in that they produce the same product (milk or meat) using the same species of livestock or close relatives (bovines). Finally, it is necessary to normalize the environmental effects of both the systems on unit output basis for a valid comparison.

Following this approach we have made an attempt in this chapter to compare the milk production systems of India and the United States for their environmental effects. As one would expect there are notable contrasts between the two systems. Most of the cattle population in the US is geared to beef production. And, with far fewer number of milk cows, the US produces a large volume of milk output. In 1992, for which we have comparable statistics, with just 9.7 million milk cows, the US produced as much as 68.6 million tonnes of milk. And, India, with its vast army of breeding cows and buffaloes (57 million cows and 40 million buffaloes), produced only 55 million tonnes of milk. It is not surprising that the milk yield of an average US cow (19.40 kg per day) is more than nine times that of an average Indian cow in-milk (2.11 kg per day), and six times that of a buffalo in-milk (3.28 kg per day). Besides being a highly selected stock, most US cows are of Holstein–Friezian breed, and less than 10 per cent are said to be dry during the year. Most Indian cows and buffaloes are said to be 'non-descript', and as many as 45 per cent of both the groups remain dry. The quality and level of feeding, as one would expect, is vastly different. We do not have feed consumed by a US cow on raw material basis. But her daily feed intake is over 20 kg in 'equivalent feeding value of corn', 56 per cent of which is in the form of concentrates (feed-grains and meals), that is 11.23 kg per day. The poor average Indian cow in-milk gets no more than 273 gm, and buffalo in-milk 947 gm of concentrates a day. Roughages form the bulk of their daily feed intake—10 kg for a cow in-milk and 15 kg for a buffalo in-milk—40 to 50 per cent of which consists of dry fodder, that is crop residues. This is the weakness as well as strength of the Indian milk production system: weakness, because it provides for low milk yield; strength, because that vast army of milch animals, probably so required, operates as a gigantic machine which transforms roughages, about half of which are crop residues, into milk. In this context it is notable that for feeding its high yielding milk cows, while the US uses about 3.6 hectare of land per cow per year (about 0.5 hectare for producing feed-grains, another 0.5 for silage and hay, and over 2.5 hectare for pasture grazing), India rests satisfied by allocating a mere 3 cent of a hectare per cow in-milk and about 6 cent of a hectare per buffalo in-milk per year for producing green fodder consumed by them respectively.

Hence, with poor quality feeding and a vast number of milch animals, it is to be expected that India's milk production system will produce very much more methane emission than the US milk production system. Indeed that is the case. In 1992, while methane emission from the Indian milk production system was as much as 4.216 million tonnes, from the US system it was only 1.747 million tonnes, even though an average US milk cow produced 180 kg of methane emission, that is about five times as much as an Indian milch cow (36 kg) and over three times as much as an Indian milch buffalo (53 kg). But methane emission per tonne of milk produced was only 26 kg by the US system as against 83 kg by the Indian bovine milk production system as a whole.

However, at the base of the real natural resource, namely land deployed for milk production in the two countries, the higher efficiency of the US system in terms of the environment becomes questionable. It turns out that in 1992 India's bovine milk production system produced over 14 tonnes of milk per hectare of land deployed, and produced 83 kg of methane emission, as against only about 2 tonnes of milk per hectare of land, and 26 kg of methane per ton of milk produced by the US system. There is apparently an indeterminacy about the environment related efficiency of the two systems. The Indian system produces about seven times more milk than the US system per hectare of land used. But, at the same it produces about four times as much methane emission per tonne of milk produced. There is, thus, a trade-off between methane emission and land used for milk production.

The local environmental effects of the Indian milk production system are positive and substantial. By recycling agricultural by-products as feed India saves a considerable amount of a natural resource, namely land. The US uses a considerable amount of land for producing milk, over half a hectare per tonne of milk produced. Manure output from the Indian milk production system has a positive effect on the local environment. Used as organic fertilizer, according to our estimate for 1992, it replaced 1.7 million tonnes of chemical fertilizers, and used as domestic fuel it replaced an amount of fuel-wood which would, otherwise, have required 900 thousand ha of land. On the US, side droppings of the milk cows on pasture lands are deemed to have positive environmental effects. But manure output at large production facilities, whether stored in lagoons and slurry-pits, or spread daily on the surrounding fields, is known to have polluting effects, such as soil contamination, nitrate pollution of the underground and surface water, and release of noxious gases to the atmosphere.

Finally, a few observations are in order for those who make country-wise comparison of methane emission from their livestock sources. As we said at the outset, such gross comparisons are neither fair nor scientifically correct. When one takes account of the differences in the human populations of the countries under comparison, things may turn out to be different. For example, in 1992, the total methane emission from India's bovine milk production system (from enteric fermentation and manure management) was more than twice (4.216 million tonnes) the emission from the US milk production system (1.747 million tonnes). But emission per capita from the Indian system was much less (5 kg) than from the US system (6.8 kg). Similarly, in the same year, methane emission from the whole of India's bovine livestock production system was 9.44 million tonnes as against 6.48 million tonnes from the cattle production system of the United States. But in per capita terms India produced less than half (11 kg) the methane emission produced by the United States (25 kg). Methane emission from livestock sources is supposed to be anthropogenic. Normalization of the total methane emission to a common anthropogenic base, thus, puts India at a level lower than the United States.

APPENDIX TABLE 10.1: US LIVESTOCK POPULATION, 1992 AND 1996

S. No.	Class	Population (1992)	% of total	Population (1996)	% of total	% change in 1996 over 1992
A	Milk cows	9.728	5.57	9.416	5.32	-3.20
(i)	Replacements heifers (226 kg and above)	4.131	2.36	4.104	2.32	-0.55
(ii)	Replacements heifers (<226 kg)	4.131	2.36	4.104	2.32	-0.55
	Total dairy cattle	17.990	10.31	17.624	9.97	-2.03
B	Beef cows	33.007	18.92	35.228	19.93	6.72
(i)	Replacements heifers (226 kg and above)	5.643	3.23	6.179	3.49	9.49
(ii)	Replacements heifers (<226 kg)	5.643	3.23	6.179	3.49	9.49
(iii)	Heifers (226 kg and above)	8.048	4.61	9.946	5.62	23.58
(iv)	Steers (226 kg and above)	16.424	9.41	17.732	10.03	7.96
(v)	Bulls (226 kg and above)	2.239	1.28	2.392	1.35	6.83
(vi)	Males (under 226 kg)	8.562	4.90	8.205	4.64	-4.16
	Total beef cattle	79.556	45.62	85.861	48.58	7.91
C	All cattle and calves	97.556	55.93	103.485	58.56	6.07
D	Others					
(i)	Sheep and lambs	10.797	6.19	8.461	4.78	-21.63
(ii)	Goats	2.000	1.14	2.545	1.44	27.25
(iii)	Hogs and pigs	58.202	33.37	56.171	31.78	-3.48
(iv)	Horses	5.85	3.35	6.050	3.42	3.41
	Total livestock	174.405	100.00	176.714	100.00	1.25
	poultry	1642.365		2096.618		27.65

Source: USDA, *Agricultural Statistics*, 1997.

APPENDIX TABLE 10.2: AREA UNDER ANIMAL FEED-GRAIN, HARVESTED ROUGHAGES, AND PASTURE LAND IN THE US, 1992

S. No.	Items	Area (million hectare)	% of total
A	Feed-grains	24.190	19.42
1	Corn	19.280	7.51
2	Sorghum	2.430	0.94
3	Oat	1.150	0.44
4	Barley	1.330	0.51
B	Harvested roughages	24.371	9.49
C	Pasture land	208.022	81.07
	Total land under animal feed production	256.583	100.00

Sources: *Agriculture Statistics, Grain and Feed*, 1994, NASS, USDA; *1997 Census of Agriculture*, United States Data, NASS, USDA.

APPENDIX TABLE 10.3: US ANIMAL UNITS

S. No.	Animal Type	GCAU	HPAU	RCAU	G & RCAU
A	Cattle				
1	Dairy				
(*i*)	Cows	1.0475	1.0000	1.0354	1.0397
(*ii*)	Heifers	0.1761	1.3038	0.8150	0.5662
2	Beef				
(*i*)	Cattle on feed	1.5323	0.5801	0.1585	0.6860
(*ii*)	Others	0.0547	1.7216	0.7358	0.4729
B	Sheep	0.0194	4.9195	0.1715	0.1078
C	Goats	0.2301	0.0611	-	-
D	Horses and mules	0.2043	-	0.3953	0.3197
E	Poultry				
(*i*)	Layers	0.0217	1.5896	0.0008	0.0090
(*ii*)	Broilers	0.0020	2.7932	-	0.0007
(*iii*)	Pullets	0.0054	3.3596	-	0.0021
(*iv*)	Turkey	0.0155	3.5520	0.0011	0.0074
F	Hogs	0.2285	1.2703	0.0294	0.1064

Notes: GCAU: Grain Consuming Animal Unit
HPAU: High-protein Animal Unit
RCAU: Roughage Consuming Animal Unit
G & RCAU: Grain and Roughage Consuming Animal Unit

Source: Baker, ERS, USDA (personal communication).

APPENDIX TABLE 10.4A: LAND UNDER FEED-GRAIN AND ITS DISTRIBUTION AMONG DIFFERENT CATEGORIES OF LIVESTOCK IN THE US, 1992

S. No.	Animal type	Population (million)	Scale factor	GCAUs (million)	% of total	Distribution of grain consumed in equivalent feeding value of corn	Per unit (tonnes/ year)	Per animal (tonnes/ year)	Distribution of land under feed-grain production (m. ha)	Per Unit (ha)	Per animal (ha/ year)
A	Cattle										
1	Dairy										
(i)	Cows	9.728	1.0475	10.19	18.92	33.06	3.24	3.39	4.57	0.45	0.46
(ii)	Heifers	8.262	0.1761	1.45	2.69	4.70		0.56	0.65		0.07
2	Beef										
(i)	Cattle on feed	10.22	1.5323	15.66	29.08	50.82		4.97	7.03		0.68
(ii)	Others	69.676	0.0547	3.81	7.07	12.35		0.17	1.71		0.02
B	Sheep	10.797	0.0194	0.21	0.39	0.68		0.06	0.09		0.01
C	Goats	2.000	-	-	-	-		-	-		-
D	Horses and Mules	5.850	0.2043	1.19	2.21	3.86		0.65	0.53		0.09
E	Poultry										
(i)	Layers	121.103	0.0217	2.63	4.88	8.52		0.07	1.18		0.009
(ii)	Broilers	1164.089	0.002	2.33	4.32	7.54		0.006	1.04		0.001
(iii)	Pullets	243.267	0.0054	1.31	2.43	4.24		0.01	0.58		0.002
(iv)	Turkey	114.426	0.0155	1.77	3.28	5.73		0.05	0.79		0.006
F	Hogs (pigs)	58.202	0.2285	13.29	24.68	43.13		0.74	5.97		0.102
	Total			53.84	100.00	174.76			24.19		

Notes: Cattle on feed are 12.85 of total beef cattle; Concentrates excludes meal, i.e. 43.27 million tonnes; Concentrates are expressed in equivalent feeding value of corn.

Sources: Allen J. Baker, ERS, USDA, 2001 (USDA, personal communication); *Agricultural Statistics*, 1997; USDA, *Agricultural Statistics*, 1994.

APPENDIX TABLE 10.4B: ROUGHAGE CONSUMING ANIMAL UNITS (RCAUs)

S. No.	Animal type	Population (million)	Scale factor	RCAUs (million)	% of total	Distribution of roughage consumed in equivalent feeding value of corn (MT)	Per unit (tonnes/ year)	Per animal (tonnes/ year)	Distribution of land under feed-grain production (m. ha)	Per unit (ha)	Per animal (ha/ year)
A	Cattle										
1	Dairy										
(i)	Cows	9.728	1.0354	10.07	13.27	31.29	3.10	3.21	30.83	3.06	3.16
(ii)	Heifers	8.262	0.815	6.73	8.86	20.89		2.52	20.58		2.49
2	Beef										
(i)	Cattle on feed	10.22	0.1585	1.61	2.12	5.00		0.48	4.92		0.48
(ii)	Others	69.676	0.7358	51.26	67.55	159.33		2.28	156.97		2.25
B	Sheep	10.797	0.1715	1.85	2.43	5.73		0.53	5.64		0.52
C	Goats	2.000	0.0611	0.12	0.16	0.37		0.18	0.37		0.18
D	Horses and Mules	5.850	0.3953	2.31	3.04	7.17		1.22	7.06		1.20
E	Poultry										
(i)	Layers	121.103	0.0008	0.09	0.12	0.28		0.002	0.27		0.002
(ii)	Broilers	1164.09	-	-	-	-		-	-		-
(iii)	Pullets	243.267	-	-	-	-		-	-		-
(iv)	Turkey	114.426	0.0011	0.125	0.16	0.37		0.003	0.37		0.003
F	Hogs (pigs)	58.202	0.0294	1.71	2.25	5.30		0.09	5.22		0.08
	Total			75.875	100.00	235.87			232.39		

Source: Appendix Table 10.4a.

APPENDIX TABLE 10.4C: GRAIN AND ROUGHAGE CONSUMING ANIMAL UNITS (G & RCAUs)

S. No.	Animal type	Population (million)	Scale factor	G&RCAUs (million)	% of total	Distribution of grain consumed in equivalent feeding value of corn (MT)	Per unit (tonnes/ year)	Per animal (tonnes/ year)	Distribution of land under feed-grain production (m. ha)	Per unit (ha)	Per animal (ha/ year)
A	Cattle										
1	Dairy										
(i)	Cows	9.728	1.0397	10.11	15.04	61.75	6.10	6.34	38.59	3.81	3.96
(ii)	Heifers	8.262	0.5662	4.67	6.95	28.53		3.45	17.83		2.15
2	Beef										
(i)	Cattle on feed	10.22	0.686	7.01	10.43	42.82		4.18	26.76		2.61
(ii)	Others	69.676	0.4729	32.94	49.02	201.29		2.88	125.77		1.81
B	Sheep	10.797	0.1078	1.16	1.72	7.06		0.65	4.41		0.41
C	Goats	2.000	-	-	-	-		-	-		-
D	Horses and mules	5.850	0.3197	1.87	2.78	11.41		1.95	7.13		1.21
E	Poultry										
(i)	Layers	121.103	0.009	1.08	1.6	6.57		0.05	4.1		0.03
(ii)	Broilers	1164.09	0.0007	0.81	1.2	4.92		0.004	3.07		0.002
(iii)	Pullets	243.267	0.0021	0.51	0.75	3.07		0.012	1.92		0.007
(iv)	Turkey	114.426	0.0074	0.84	1.25	5.13		0.04	3.20		0.027
F	Hogs (pigs)	58.202	0.1064	6.19	9.21	37.81		0.64	23.63		0.406
	Total			53.84	100.00	410.63			256.43		

Source: Appendix Table 10.4a.

APPENDIX TABLE 10.4D: DISTRIBUTION OF MEAL AMONG GRAIN CONSUMING ANIMAL UNITS (GCAUs)

S. No.	Animal Type	Population (million)	GCAUs (million)	% of Total	Meal Distribution (MT)	Meal per unit/ year (tonnes)	Meal per animal/ year (tonnes)
A	Cattle						
1	Dairy						
(i)	Cows	9.728	10.19	18.92	6.92	0.68	0.71
(ii)	Heifers	8.262	1.45	2.69	0.98		0.11
2	Beef						
(i)	Cattle on feed	10.22	15.66	29.08	10.64		1.04
(ii)	Others	69.676	3.81	7.07	2.58		0.04
B	Sheep	10.797	0.21	0.39	0.14		0.01
C	Goats	2.000	-	-	-		-
D	Horses and mules	5.850	1.19	2.21	0.8		0.13
E	Poultry						0.01
(i)	Layers	121.103	2.63	4.88	1.78		0.001
(ii)	Broilers	1164.089	2.33	4.32	1.58		0.003
(iii)	Pullets	243.267	1.31	2.45	0.88		0.010
(iv)	Turkey	114.426	1.77	3.28	1.2		0.010
F	Hogs (pigs)	58.202	13.29	24.68	9.3		0.155
	Total		53.84	100.00	36.62		

Note: The quantity of meal has been converted into million metric tonnes (one million metric tonne = 1.1023 short million tons); we have assumed that the quantity of meal expressed in this table is equivalent feeding value of corn.

Source: Appendix Table 10.4a.

APPENDIX TABLE 10.5: PER ANIMAL FEED CONSUMPTION AND LAND DEPLOYED BY US CATTLE, HOGS, AND POULTRY, 1992

S. No.	Animal type	Feed consumption per animal /per day (kg/animal/day)				Land per animal per year (hectare/ animal/year		
		Feed-grain	By-produce meal	Roughages	Total	Feed-grain	Roughages	Total
A	Cattle							
1	Dairy							
(*i*)	Cows							
(*ii*)	Heifers	9.29	1.94	8.79	20.02	0.46	3.16	3.62
2	Beef	1.53	0.30	6.70	8.73	0.07	2.49	2.56
(*i*)	Cattle on feed	13.62	2.85	1.31	17.78	0.68	0.48	1.16
(*ii*)	Others	0.47	0.11	6.25	6.83	0.02	2.25	2.27
B	Poultry							
(*i*)	Layers	0.19	0.00	0.00	0.19	0.009	0.002	0.011
(*ii*)	Broilers	0.02	0.01	-	0.03	0.001	-	0.001
C	Hogs (pigs)	2.03	0.41	0.25	2.69	0.102	0.08	0.182

Sources: USDA, *Agricultural Statistics, 1992. 1997 Census of Agriculture*—State Data, USDA, NASS.

APPENDIX TABLE 10.6: FEED AND RELATED INPUTS AND MILK OUTPUT: INDIA AND THE US, 1992

S. No.	Particulars	India	United States
I	Milk output (MT)		
(*a*)	Cow milk	24.338	68.554
(*b*)	Buffalo milk	31.044	-
II	Feed consumed (MT)		
(*a*)	By cows (milch)	3.88	-
1	Concentrates	-	33.06*
(*i*)	Feed-grains	-	6.92*
(*ii*)	By-product meals		
2	Roughages		31.29
(*i*)	Crop residues (dry fodder)	99.63	-
(*ii*)	Green fodder	96.69	-
(*b*)	By buffaloes (milch)		
1	Concentrates	10.16	
(*i*)	Feed-grains	-	
(*ii*)	By-products meals	-	
2	Roughages		
(*i*)	Crop residues (Dry fodder)	88.01	
(*ii*)	Green fodder	117.21	
(*c*)	By milch bovines (MT)	14.04	
1	Concentrates	-	
(*i*)	Feed-grains	-	
(*ii*)	By-products meal		
2	Roughages	187.68	58.52
(*i*)	Crop residues (dry fodder)	213.9	-
(*ii*)	Green fodder		-
III	Total digestible nutrients (TDN) consumed	65.99	
(*a*)	By cows	73.06	
(*b*)	By buffaloes	139.05	-
(*c*)	By milch bovines		4.57
IV	Land deployed for producing the feed consumed (million ha)		-
(*a*)	By cows	-	
1	Concentrates	-	-
(*i*)	Feed-grains	-	30.83
(*ii*)	By-product meal		-
2	Roughages	-	
(*i*)	Crop residues (dry fodder)	1.76	
(*ii*)	Green fodder (silage and hay)	-	
(*iii*)	Pasture Land		
(*b*)	By buffaloes	-	
1	Concentrates	-	
(*i*)	Feed-grains	-	
(*ii*)	By product meal		

(contd.)

S. No.	Particulars	India	United States
2	Roughages	-	
(*i*)	Crop residues (dry fodder)	2.13	
(*ii*)	Green fodder (silage and hay)	-	
(*iii*)	Pasture land		
(*c*)	By milch bovines (MT)	-	
1	Concentrates	-	
(*i*)	Feed grains	-	
(*ii*)	By-products meal		
2	Roughages	-	
(*i*)	Crop residues (dry fodder)	3.89	
(*ii*)	Green fodder (silage and hay)	-	
(*iii*)	Pasture land		

Notes: * Quality expressed in equivalent feeding value of corn.
+ Estimates for breeding cows buffaloes and bovines
** Sum of land engaged in production of harvested roughages and pasture land.

Source: See Table 10.1 of Chapter 10.

APPENDIX TABLE 10.7: LAND ENGAGED WITH GREEN FODDER PRODUCTION AND ITS DISTRIBUTION AMONG DIFFERENT CATEGORIES OF BOVINES IN INDIA, 1992

S. No.	Categories	Consumption (tonnes/head/year)	Population (MT)	Total consumption (MT)	% of total	Land engaged with categories (million hectare)	Land per animal (hectare)
1	Cow in-milk	1.857	31.562	58.61	12.76	1.07	0.034
2	Dry cow	1.445	26.231	37.90	8.25	0.69	0.026
3	Adult male cattle	2.153	74.489	160.37	34.90	2.93	0.039
4	Cattle young stock	0.649	65.734	42.66	9.29	0.78	0.012
	Total cattle		198.01	299.54	65.19	5.48	0.028
5	Buffalo in-milk	3.201	25.892	82.88	18.04	1.53	0.058
6	Dry buffalo	2.387	14.383	34.33	7.47	0.62	0.043
7	Adult male buffalo	2.376	8.086	19.21	4.18	0.35	0.043
8	Buffalo young stock	0.726	32.314	23.46	5.11	0.42	0.013
	Total buffalo		80.675	159.88	34.80	2.92	0.036
	Total		278.685	459.42			0.030

Note: The population of dry females, cattle as well as buffaloes, is exclusive of the economically useless category called 'others' and also adult females, (> 3 years of age) 'not calved even once'.

Sources: Consumption data from Jain and Singh (1990); Land under green fodder production from GOI, *Basic Animal Husbandry Statistics, 1997.*

APPENDIX TABLE 10.8: FUEL USED IN CORN PRODUCTION, STATE-WISE

S. No.	States	Diesel		Gasoline		L.P. GAS	
		Gallons per acre	Kg. per hectare	Gallons per acre	Kg. per hectare	Gallons per acre	Kg. per hectare
1	Colorado	8.36	78.15	3.93	36.75	-	-
2	Illinois	4.46	41.69	3.51	32.83	2.11	19.74
3	Indiana	5.11	47.77	3.56	33.27	2.12	19.81
4	Iowa	4.44	41.52	3.41	31.89	4.67	43.67
5	Michigan	6.9	54.64	3.1	28.97	3.18	29.74
6	Minnesota	4.72	44.11	2.88	26.92	4.28	40.01
7	Nebraska	17.89	167.24	4.47	41.79	3.56	33.27
8	Ohio	4.5	42.06	2.62	24.50	3.65	34.14
9	South Dacota	6.11	57.13	3.05	28.50	4.9	45.82
10	Wisconsin	7.62	71.23	2.55	23.83	1.96	18.33

Source: *Corn: State-level Production Costs, Characteristics, Inputs and Machine Use Data*, ERS, USDA, 1991.

11. Summary and Conclusions

11.1 Introductory Observations

In the present study an attempt has been made to comprehend the environmental dimensions of India's livestock production system, and to measure and evaluate the effects of the system on the environment. We have also made an attempt to compare India's milk production system with that of the United States of America for their environmental effects, with the intention of seeing how the system of a developed country stands *vis-à-vis* that of a developing country like India insofar as its environmental effects are concerned. In the contemporary international concern about greenhouse gases emission, livestock have been singled out as the most important source of methane emission after rice cultivation. And, India, due to its having the largest livestock holding in the world, has been singled out for the largest amount of methane emission from livestock sources. Concern about methane emission is well placed, of course, since methane is found to have many times more global warming potential (GWP) than the ubiquitous carbon dioxide. But a country-wise comparison of gross amounts of methane emissions, we have contended, is neither fair nor scientifically correct. For a fair and correct comparison it is required that not only methane emission but any other environmental effects of two different livestock production systems be first normalized, in other words reduced to a common base, such as per capita of human population or per unit of output, and then compared. Further, an exclusive focus on methane emission leads to a livestock production system(s) not receiving a proper comprehensive treatment of its environmental effects, which occur both on the output as well as input side of the system, and which, depending upon their nature, impact upon the global or local environment, directly or indirectly. Output of methane, for instance, causes global warming. Output of dung used as organic manure has a double-effect. It replaces equivalent amount of chemical fertilizers, and preserves the natural fertility of the soil, a local environmental effect. At the same time it prevents greenhouse gases emission that would otherwise occur in the manufacture, distribution, and supply of the chemical fertilizers at the farm gate level, an indirect global environmental effect. On the input side, production of feed-grains consumed by animals, as in the US system, causes greenhouse gases emission from the fossil fuel used in the process, besides using land, a natural resource. Agricultural by-products used as animal feed, as in India, on the other hand, saves land, an all-time limitational natural resource. Rather than a piecemeal, pick and choose approach, a systems' approach to the study of environmental effects of livestock production is the right one. Such an approach has been followed in the present study.

At the outset, two points are worth noting. First, livestock are a part of India's natural resource endowment. Domesticated in the prehistoric times, they have adapted to the highly diverse environmental conditions of the country, which has almost every type of climate prevalent. This diversity has determined the biodiversity of the stock and the genetic variability of different species. Scientific knowledge about these characteristics is still very scarce. What is known, however, is that post-independence livestock breeding policies, guided as these were by considerations of short-term productivity gains, have not been helpful to the preservation of livestock biodiversity and genetic variability. Second, the livestock play an important role in the economic life of the Indian people. The livestock sector's share in the national gross domestic product (GDP) has remained constant over the years at between 5 to 6 per cent of the total, while the share of 'agriculture and allied activities', of which it forms a part, has been declining. Within the latter, livestock's share has been increasing, however, as the relative importance of animal food *vis-à-vis* food grains has been increasing with growing per capita incomes in the country. Its share in the value of output of 'agriculture and allied activities' at 1980–1 prices increased from 19 per cent in 1980–1 to 23 per cent in 1993–4. Similarly, its share in the latter sector's GDP increased from about 13 per cent in 1980–1 to 19 per cent in 1993–4. The livestock are a significant means of livelihood for millions of people, especially the poorer sections in rural India. Assuming the same labour to output ratio as in agriculture, the livestock sector currently provides over 43 million person-years of work annually.

At the base of the livestock economy is a large livestock population. In 1992, the latest year for which census figures are available, India had a livestock population of 471 million heads. Cattle and buffaloes (bovines) were largest in number (289 million) followed by small ruminants, goat (115 million), and sheep (51 million). This vast number is deployed in the production of animal food (milk, meat), fibre (wool), and draught power. A number of studies since the 1950s have shown that feed, fodder availability in the country has been less than the nutritional requirement of the population. Productivity of animals is, therefore, less than the potential. Although, in number bovines account for 61 per cent of the total livestock, in terms of 'cow-units', based on feed, fodder consumption, they account for 90 per cent of the total. In view of their overwhelming importance we have focused attention on the environmental effects of the bovine production system. Livestock and bovine are terms which we have, for this reason, often used synonymously.

Several direct and indirect environmental effects both on the input as well as output side of the cattle and buffalo based production systems have been identified. These fall into two broad groups: *global* and *local*. The global effects are greenhouse gases' emission actually realized or prevented. Specifically, these are: (*i*) methane emission from enteric fermentation and 'manure management'; (*ii*) methane emission from burning of dung-cake as domestic fuel; and (*iii*) prevention of CO_2 emission associated with the manufacture, distribution, and supply of chemical fertilizers equivalent to the amount of dung used as organic manure. The identified local environmental effects are: (*i*) land saving due to agricultural by-products used as animal feed; (*ii*) land saving due to dung used as biomass fuel in place of fuel-wood requiring land for production; and (*iii*) saving of chemical fertilizers due to dung used as organic manure.

In order to quantify the above effects, relationships linking the effects with the relevant input or output of the production system have been incorporated into a mathematical model. It is a simple static model having a modular structure, a module each for the environmental effects enumerated earlier. A module can be detached from the model and used separately for computational purposes, provided one has the data on the variables and parameters appearing in the module The model can be used for simulation of the system under varying conditions of production, including due to alternative policy interventions, as well as for checking uncertainty associated with the data inputs. It can also be dynamized by linking it with a cattle and buffalo population generation sub-model. A summary of our own computational results from the model is given in the following sections. Note that all data inputs and, therefore, the results relate to the year 1992.

11.2 Global Environmental Effects

The computational results show that the total methane emission from the livestock production system in 1992 was 9.767 million tonnes. Emission from 'manure management' (0.202 million tonnes) and burning of dung cake as domestic fuel (0.21 million tonnes) are insignificant in comparison to emission from enteric fermentation (9.355 million tonnes). The reason is that dung output is India in handled in solid form and stored in the open, aerobic condition such that it provides little or no opportunity to methanogenic bacteria. And, methane emission from burning of dung cake is a trace gas any way. We have accounted it separately and in addition to methane emission from 'manure management', which includes dung output used for making dung cake. A notable point about our estimate of methane emission from enteric fermentation is that it is based upon gross energy *intake* derived from feed consumption rates estimated from primary data collected through sample surveys. Compared to other estimates based upon estimated gross energy *requirement* of the animals, our estimate involves fewer assumptions and more of primary data, and hence, should be more reliable. In the total methane emission from enteric fermentation, cattle's share (68 per cent) is marginally lower than its share in the bovine population (71 per cent) while buffaloes's share (32 per cent) is marginally higher than its share in the population (29 per cent). Interestingly, whereas within buffaloes, in-milk buffaloes account for close to half of methane emission from enteric fermentation, within cattle, adult males, most of them being working bullocks, account for about half of the emission.

From the point of view of methane emission reduction it is usual to focus on dairy animals. In terms of a set of two environment related efficiency criteria: methane emission per unit of milk produced, and milk produced per unit of energy intake, the buffalo based dairy production system turns out to be more efficient than the cow based system. The former produces more milk per unit of gross energy intake and less methane emission per unit of milk produced than the cow based dairy production system. Specifically, per mega joule of energy intake, a buffalo in-milk produces more milk (23 gm) compared to a cow in-milk (21 gm). At same time it produces less methane per kg of milk produced (49 gm) compared to its cow counterpart (51 gm). The same position holds when we consider the milch animals (in-milk + dry) as a whole. Milk produced per mega joule of energy intake per milch buffalo works out to be larger (16 gm) than per milch cow (13 gm). And, methane emission per kg of milk produced is lower (69 gm) compared to the cow based system (86 gm). This sort of relative efficiency is further confirmed when instead of gross energy intake one looks at milk produced per unit of total digestible nutrients (TDN) consumed. While the pattern of methane emission per unit of milk produced remains as it is, milk produce per unit of TDN consumed by milch buffalo is larger (424 gm per kg TDN) than by milch cow (367 gm per kg TDN). The choice between the two system is not unconstrained, however. The cow based production system's primary function, unlike that of buffalo based system, is to produce and reproduce working bullocks.

The environmental value of animal energy use in Indian agriculture, especially by tens of millions of small and marginal farmers, has hardly been realized. This is a renewable and sustainable form of bio-energy. The working animals' population derives its energy largely from by-products of agriculture, whose social opportunity cost is zero. Its alternative, fossil fuel energy like diesel, is unsustainable and costly, and burning it causes greenhouse gases emission, in particular CO_2 emission.

The agricultural engineers and energy experts who seek equivalence between working animals and tractors in terms of horsepower make the mistake of assuming their normative equivalence to be the same as the economic rate of substitution between the two. The real rate of substitution, on the contrary, is determined by a host of economic factors, such as land use pattern, farm-size and landholding structure, cropping patterns, etc. Micro farm level studies have found that only three to four working animals are replaced by tractor in real life. Our own investigation at the state level, has confirmed those findings. At the all-India level, our conclusion is that a tractor replaces not more than 12 working animals. At this rate, if more than 70 million working animals (1992) were to be replaced by tractors, approximately 6 million tractors would have to be in place to carry on agricultural production in India. Anually it will cost about 20 million tonnes of diesel, the burning of which in running the tractors, according to our estimate, would cause over 6 million tonnes of carbon dioxide emission that gets prevented annually by animal energy use in agriculture. It is equivalent to 0.30 million tonnes of methane, methane's global warming potential (GWP) being 21 times higher than that of carbon dioxide. Taking credit for this amount of methane emission prevented, total methane emission from the livestock system works out to 9.467 million tonnes in 1992. We have not, for want of data, been able to account for carbon dioxide emission prevented due to dung used as manure in place of chemical fertilizers. Once we are able to take credit for that, we expect the methane balance of the country's livestock production system to be around 9 million tonnes in 1992.

11.3 Local Environmental Effects

As noted earlier, India's livestock production system saves land in two ways: (*i*) by using agricultural by-products as animal feed, and (*ii*) by using a part of the dung output as domestic fuel. Any activity that saves or displaces a natural resource is a nature conserving, environment protecting activity. The land saving on account of by-products feed is considerably large. If the by-products feed were to be replaced by green fodder, as much as 1,449 million tonnes of the latter will be required to provide the same amount of energy as from the by-products feed. And, to produce it 46.44 million ha of land will be required at the average green fodder yield of 31.08 tonnes per hectare. The land saving on account of dung cake output of 68 million tonnes used as domestic fuel is relatively small, about 2 million hectare to produce the replacement quantity of fuel wood of 19 million tonnes at the fuel wood yield of 30 tonnes per hectare, and gestation lag of 3 years between planting and harvesting of fuel wood trees. The total land saving due to the livestock production system, thus, works out to over 48 million ha in 1992. Its order of importance can be gauged by noting that it is as much as 34 per cent of the country's net sown area of 142 million ha.

The dung output used as manure on dry matter basis works out to about 103 million tonnes in 1992. If it was to be replaced by chemical fertilizers, 2.5 million tonnes of ammonium sulphate,

0.46 million tonnes of super phosphate, and 0.78 million tonnes of muriate of potash will be required. The amount of chemical fertilizers thus saved does not give the full environmental value of the dung used as manure. It is indeed the saving of mineral resources required to manufacture the substitute amounts of chemical fertilizers which ought to count for the true environmental value of dung manure. Mineral resources are exhaustible. To the extent they are used, they get depleted. Dung used as manure ultimately prevents their depletion. In addition, besides being sustainable, dung manure, unlike chemical fertilizers, preserves the natural fertility of the soil and does not harm soil-born micro-organisms.

11.4 Valuation of the Effects

The method of valuation of the environmental effects of a livestock production system, or for that matter any economic project or activity is that of *social cost–benefit analysis*. An environmental effect, depending upon the nature of its impact, is supposed to have direct or indirect social benefits or costs. Methane emission, for instance, is known to have a negative impact, causing global warming and possible climate change worldwide. However, even if one is able to quantify it, as we have done, it is not possible to put a negative value (cost) on it. The reason is that its impact can not be distributed according to the countries of its origin. Most local environmental effects of India's livestock production system, as we have seen, have a positive impact. The social value or benefit of these is truly large, as much as Rs 1,16,518 crore in 1992, much more than the value of output of the livestock sector (Rs 78,682 crore in 1997). The major portion of this is due to agricultural by-products used as animal feed (Rs 76,357 crore). This has been obtained by multiplying the land area saved by average yield of rice under rainfed conditions and the wholesale market price of rice. Next in importance is the value of animal energy used in agriculture (Rs 35,357 crore). It is equal to the value of diesel saved or replaced at the retail price of diesel in Delhi. The last in line of importance are environmental value of dung cake used as biomass fuel (Rs 3,144 crore), and dung used as organic fertilizer (Rs 1,660 crore). The latter is equal to the value of the replacement amounts of chemical fertilizers at current prices. And, the former has been estimated in the same manner in which the value of by-products feed has been estimated that is the land area saved multiplied by the rice yield, and the wholesale price of rice. Few could have imagined that the social value of the positive environmental contribution of India's livestock production system could be so large until it has been demonstrated. A note of caution, however, is necessary. The estimates given above should be treated as orders of magnitude involved rather than precise numbers.

11.5 Comparison with the United States

The idea behind the comparison has been to see where does India, a developing country, stand *vis-à-vis* a developed country like the United States in terms of the environmental effects of their livestock production systems. The comparison is, however, confined to what is comparable, that is the environmental effects of the milk production systems of the two countries. With far fewer number of milk cows, the US produces a large volume of milk output. In 1992, with just over 9.7 million cows, the US produced about 69 million tonnes of milk. And, India with it vast army of breeding cows and buffaloes (57 million cows + 40 million buffaloes in-milk and dry) produced only 55 million tonnes of milk. It is not surprising that the milk yield of an average US cow (19.4 kg per day) is more than nine times that of an average Indian cow in-milk (2.11 kg per day) and six times that of cow in-milk (3.28 kg per day). Differences in the quality (breed-type) of animals apart, the feed intake of a US cow is over 20 kg a day in 'equivalent feeding value of corn', 56 per cent of which comes from concentrates (feed-grains and meal). With marginal supplementation of concentrates, feed intake of Indian cows and buffaloes is in the form of roughages—10 kg a day for cow and 15 kg a day for buffalo in-milk on raw material basis, 40 to 50 per cent of which is in the form of dry crop residues. Thus, the Indian army of milch bovines makes milk out of roughages, although at low milk yields. For feeding its high yielding milk cows the US uses 3.6 hectare of land per cow per year, about half hectare for producing feed-grains consumed, another half hectare for silage and hay, and over 2.5 hectare for pasture grazing. India, on the other hand, remains satisfied by allocating a mere 3 cent of a hectare per cow, and 6 cent of a hectare per buffalo in-milk for producing green fodder supplement consumed by them respectively (see Appendix Table 10.7)

Against the above sharply contrasting patterns of milk production, quality and quantity of feeding and land resource use, an attempt has been made in this study to compare the local and global environmental effects of the two milk production systems. Consider the local effects first. We have already seen how important and positive are the local effects from by-products feeding and dung used either as organic manure or as biomass fuel in India. We can separate out these effects that are exclusively due to the milk production system. But there are no comparable effects on the US side. All that is known is that the US instead of saving land uses a considerable amount of land for producing milk, over half a hectare per tonne of milk produced. At the rate of 41 kg a day per milk cow, it produces a large volume of dung annually, 145 million tonnes in 1992 on wet basis. The amount of dung dropped on the pastures is expected to have regenerative effect on the pasture biomass. Most of the dung output seems to be stored in lagoons or slurry-pits, and some possibly spread daily on the fields surrounding the large production facilities. Either way its local environmental effects are reported to be harmful, the notable effects, being soil contamination, nitrate pollution of the underground and surface water, release of noxious gases to the atmosphere, and damage to aquatic life.

As for the global effects, that is methane emission from enteric fermentation and 'manure mangaement', the comparison of the two systems is more specific and concrete. An average US milk cow produces 180 kg of methane a year (see USEPA 1998, Annex H&T, Draft Report), about five times as much as an average Indian milch cow (36 kg), and over three times as much as an Indian milch buffalo (53 kg). Yet, because of the army of milch animals, 45 per cent of which remain dry during the year, the total methane emission from the Indian milk production system was far more (4.216 million tonnes) than from the US milk production system (1.747 million tonnes) in 1992.

When methane emission is compared on per animal basis it does not speak about the environment related efficiency of the two systems. The latter is obtained when one compares milk output per unit of feed intake and methane emission per unit of milk produced. Since there are difficulties in comparing feed intakes per unit of milk output, particularly because we do not have information on different feeds consumed on the US side on comparable raw material basis, we have taken the land used for feeding the animals on either side to produce a unit output of milk. Moreover, land is the real natural resource which is at the base of feed input, and feeding practices. At the base of land used in milk production, India scores over the United States in terms of milk production. In 1992, the Indian system produced over 14 tonnes of milk per hectare of land used as against 2 tonnes by the US system. But, at the same time the Indian system produced as much as 83 kg of methane emission per tonne of milk produced, while the US system produced only 26 kg. Apparently, there is, thus, a trade-off between methane emission and land used for milk production. On the one side you use more land per unit of milk produced and produce less methane emission, on the other you use less land and produce more methane emission. There is, thus, an indeterminacy about the environment related efficiency of the Indian and US milk production systems.

Lastly, it may be of interest to those who make country-wise comparisons of gross amounts of methane emission from livestock sources to know that per capita of human population the Indian livestock production system emits less methane than the US system. In 1992, methane emission per capita from India's milk production was less (5 kg) than that from the US system (6.8 kg). Similarly, methane emission per capita from the whole of India's bovine production system was less than half (11 kg) of the emission (25 kg) from the US cattle production system. Methane emission from the two livestock systems when normalized on common anthropogenic base thus puts India at a level lower than the United States.

11.6 Concluding Remarks: Limitations of the Study

Finally, it is necessary to be modest about what we have achieved in this study. Being the first of its kind in which an attempt has been made to deal with the environmental dimensions of India's livestock production system, it retains an exploratory character. No claim is made that all, and each one, of the direct and indirect environmental effects of the system has been accounted for. For instance, recycling of crop residues (dry fodder) as animal feed in India besides having land saving-effect has another indirect effect which, one might suggest, remains unaccounted. It is the emission of methane as a trace gas that would otherwise occur if the crop residues were to be burnt instead of being used as animal feed. That much emission of methane, thus, gets prevented by the Indian system of using crop residues as animal feed. Were it to be credited, the methane balance from the livestock production system will get reduced. The environmental effects of feed consumed by grazing, similarly, remain unaccounted for want of data. Were it to be included it is easy to conjecture that our estimate of CH_4 emission from enteric fermentation will marginally go up. At the same time, animal-droppings on the pastures should be expected to have a regenerative effect.

Some caveats are also in order in respect of the estimates of the environmental effects which have been accounted for. There is no sanctity about the parameter values that have been used for estimation purposes. The reader may have noticed that at appropriate places in the text we have pointed out to the difficulties associated with the parameter values. Even then a critic may come up to say that the rate of methane emission from enteric fermentation used is on the lower side, or for that matter, the feed consumption rates used are lower or higher than what they ought to be. In our judgement we have used the best estimate values of the parameters available to us. Yet we concede that there is a measure of uncertainty surrounding the parameter values. That uncertainty can be handled, as we have said earlier, by a sensitivity analysis, in other words by changing the parameter values upward and downward marginally, and looking at the stability of the estimated values of the environmental effects. The best is to simulate the model (Chapter 4) under conditions of these uncertainties and produce a reasonable range of estimates of the environmental effects. The results of any empirical study always have scope for improvement.

APPENDIX

A Note on the Estimation of Dung Evacuation Rate by Different Categories of Bovines

A.1: Estimates of dung evacuation rates, i.e. dung output per day per animal according to age, sex, and functional categories of bovines are not available. However, from the Integrated Sample Surveys on livestock products and related practices carried out by the states, dung evacuation rate per bovine, though unpublished, are available with the Department of Animal Husbandry and Dairying, Ministry of Agriculture. Based on the recent estimates for several states obtained from the Statistical Advisor of the Department, we worked out an average all-India rate of about 6.10 kg dung output per day per bovine. The state level estimates are arrived at in the following manner. The field investigator, on the day of the visit to the sample household, is required to do actual weighment of the dung droppings of the household's bovine herd as a whole. This process is repeated to take care of seasonal variations. For practical reasons it has not been possible for the investigators to separate out droppings of each member of the herd during the day for weighment purposes. The evacuation rate per bovine is obtained by summing the dung output per day of the sample households, and dividing the sum by the sum of the bovine animals of the sample households as recorded on the day of the enquiry. Now, the bovine herd consists of several categories of cattle and buffaloes of different age groups, sex, and function. The average dung evacuation per day per animal will of course vary according to the age, sex, and functions of the animal.

A.2: The variation in the evacuation rates of different categories of animals is expected to be similar to the variation in their per day feed consumption, since dung output of an animal, *ceteris paribus*, is dependent on its feed consumption. Data on feed consumption per animal per day for different categories of cattle and buffaloes on all-

India basis are available from Jain and Singh (1990) (see Appendix Table 3.3). An examination of this data shows that the variation in per animal per day feed consumption has a definite pattern. Since feeds fed are of different types (concentrates, dry, and green fodder) we can see the variation in terms of total digestible nutrients (TDN) equivalents of the feeds. Among cattle the largest per day feed consumption is that of adult males, followed by cows in-milk, followed by dry cows, young males, and young females. Among buffaloes, the largest per day feed consumption is that of a buffalo in-milk, dry buffalo, followed by adult male, young-females, and males. Given these patterns of variation in feed consumption we can postulate that the variation in evacuation rates within cattle and buffaloes will also be of a similar definite pattern.

A.3: The symmetry between feed consumption rates and the evacuation rates having been assumed, we now make another assumption, which is not quite realistic but not quite unreasonable. It is that the per cattle and per buffalo evacuation rates are the same and equal to the per bovine evacuation rate. Now, given the definite pattern of variation in the per animal per day feed consumption (TDN) among different categories of cattle (buffaloes), and the two assumptions, namely; (*i*) variation in dung evacuation rates being symmetric to the variation in feed consumption rates, and (*ii*) per cattle (buffalo), dung evacuation rate being the same as per bovine dung evacuation rate, we can proceed to determine the evacuation rates of different categories of cattle (buffaloes) as follows.

A.4: Let there be n number of cattle, one each from different categories, and f_j, (j=1, 2..., n) be the feed consumption of the jth animal per day in terms of TDN. The mean feed consumption per day per cattle is then,

$$M_{fc} = \frac{\sum_{j=1}^{n} f_j}{n} \qquad \text{(A1)}$$

where M stands for mean, subscript f denotes feed, and c denotes cattle. The relative deviation of the feed consumption of the jth animal from the mean as a measure of variation is defined as:

$$p_j = \frac{f_j}{M_{fc}} \qquad \text{(A2)}$$

Following our first assumption, the evacuation rate of the jth animal will deviate from the mean evacuation rate per cattle by a factor equal to p_j such that

$$d_j = p_j M_{dc} \tag{A3}$$

where d_j denotes the evacuation rate of the *j*th animal, and M_{dc} denotes mean dung evacuation rate per cattle. Analogously, the evacuation rate of the *j*th category of buffalo is given as:

$$d'_j = p'_j M_{db} \tag{A4}$$

where M_{db} is the mean evacuation rate per buffalo. Finally, following our second assumption,

$$M_{dc} = M_{db} = M_d \tag{A5}$$

where M_d is the dung evacuation rate per bovine.

The values of d_j and d'_j can be calculated using equations (A1)–(A5). The values of f_j and M_d are given exogenously. The values of the parameters p_j and p'_j can be easily calculated using equation (A2) and its analogue for buffaloes. The results of our computation are presented in Table A1.

TABLE A1: DUNG EVACUATION RATES OF DIFFERENT CATEGORIES OF BOVINE ANIMALS

S. No.	Category of bovines	f_j in terms of TDN (kg/day)	Values of (p_j p_j)	Dung Evacuation Rate (d_j, d_j) (kg/day)
		(1)	(2)	(3)
A	Cattle			
1	Cow in-milk	3.49	1.27	7.74
2	Dry cow	2.69	0.98	5.97
3	Cattle adult male	3.56	1.29	7.86
4	Young male	2.00	0.72	4.39
5	Young female	1.98	0.72	4.39
	Mean (M_{fc})	2.744		6.11
B	Buffaloes			
1	Buffalo-in-milk	5.24	1.59	9.69
2	Dry buffalo	4.49	1.36	8.29
3	Buffalo adult male	3.11	0.94	5.73
4	Young male	1.52	0.46	2.80
5	Young female	2.12	0.64	3.90
	Mean (M_{fb})	3.29		
	Mean per bovine (kg/day)			6.1

Sources: For figures in column 1 see Appendix Table 3.3; column 2 figures calculated by us; mean evacuation rate per bovine, column 3, supplied by the Statistical Advisor, Department of Animal Husbandry and Dairying, Ministry of Agriculture, Government of India.

It may be noted that the mean feed consumption rate of the cattle group is marginally lower than that of the buffalo group. That implies that the mean dung evacuation rate of the cattle group should be marginally lower than that of the buffalo group. Since we have assumed the same mean evacuation rate for both the groups, there is marginal overestimation in the evacuation rates of the different categories of cattle, and marginal underestimation in the rates of the different categories of buffaloes. Applying the evacuation rates as given in column 3 of the table to the 1992 cattle and buffalo population yields a total of 687 million tonnes dung output. As an order of magnitude this estimate is close to the CSO's estimate of 693 million tonnes dung output in 1996–7 (obtained from DDG National Accounts' Division [CSO]), which is obtained by applying state-wise evacuation rate per bovine to the respective bovine populations. From this counter-check we conclude that our category-wise estimates of evacuation rates at the all-India level are quite reliable.

References

Amble, V.N. et al., 1965, 'Feed Requirement of Bovines and Possibilities of Meeting Them', *Indian Journal of Agricultural Economics*, vol. 20, no. 1, January–March.

Anonymous, 1959, *Domestic Fuel in India*, Survey Report, National Council of Applied Economic Research, New Delhi.

Bhattacharya, Sumana and A.P. Mitra, 1998, *Greenhouse Gas Emission in India for the Base Year 1990*. Scientific Report, no. 11, Centre for Global Change, National Physical Laboratory, Delhi.

Binswanger, Hans, 1978, 'The Economics of Tractors in South Asia: An Analytical Review', Agricultural Development Council, New York.

Burns, W., 1944, *Technological Possibilities of Agricultural Development in India*, Superintendent of Government Printing, Punjab, Lahore.

Centre for Monitoring Indian Economy, 1999, Agriculture, New Delhi.

Centrell, Patty et al., 2001, 'Environmental and Factory Farms', in *Motion Magazine*, 10 April. http://www.inmotion.magazene.com/hwenu.html.

Cooper, Richard N., 2000, 'International Approaches to Global Climate Change', *The World Bank Research Observer*, vol. 15, no. 2, August.

Crutzen, P. J. and M.O. Andrea, 1990, 'Biomass Burning in the Tropics: Impact on Atmospheric Chemistry and Biogeochemical Cycles', *Science*, vol. 250, pp. 1669–78.

de Haan, Cees, 1997, *Livestock and the Environment: Finding a Balance*, Report of Study Coordinated by the FAO of the United Nations, the United States Agency for International Development, and the World Bank.

Delmas, R.A. and D. Ahuja, 1993, 'Estimating National Methane Emissions from Anthropogenic Bio-mass Burning', in A.R. Van Amstel (ed.), Proceedings of an International IPCC Workshop: Methane and Nitrous Oxides, Methods in National Emissions Inventories and Options for Control, 3–5 February 1993, Amersfoort, NL RIVM Report, no. 481507003, Bilthoven, The Netherlands, July.

Gadgil, D.R., 1947, '*Economics of Irrigation*', Gokhale Institute of Politics and Economics, Pune.

Garg, A.C., M.A. Idnani and T.P. Abraham, 1971, *Organic Manure*, Technical Bulletin (Agriculture) 32, Indian Council of Agriculture Research, New Delhi.

Gibbs et al., 1989, *Reducing Methane Emission from Livestock: Opportunities and Issues*, Office of Air and Radiation, United States Environment Protection Agency, Washington, D.C.

Goel, B.B.P.S. et al., 1973, 'Availability and Disposal of Dung in India', *Indian Journal of Animal Science*, vol. 43, no. 8, August, pp. 671–6.

Government of India (GOI), 1976, *Report of the National Commission on Agriculture 1976, Part VII*, Ministry of Agriculture, Department of Animal Husbandry, New Delhi.

——, 1987, *Report of the Task Force to Evaluate the Impact of Sheep and Goat Rearing in Ecologically Fragile Zone 1987*, Ministry of Agriculture, Department of Agriculture and Cooperation, New Delhi.

——, 1992, *15th Indian Livestock Census, 1992*, Directorate of Economics and Statistics, Department of Agriculture and Cooperation, Ministry of Agriculture, New Delhi.

——, 1996, *Report of the Steering Group on National Livestock Policy, 1996*, Ministry of Agriculture, Department of Animal Husbandry and Dairying, New Delhi.

——, *National Accounts Statistics*, 1993 and 1996, Central Statistical Organisation (CSO).

——, 1996–7, *Agricultural Prices in India 1996–97*, Ministry of Agriculture, New Delhi.

——, *Basic Animal Husbandry Statistics*, AHS Series-7, 1999, Ministry of Agriculture, Department of Animal Husbandry and Dairying.

——, National Sample Survey Organisation (NSSO) 2000, NSS 55th Round (July 1999–June 2000), Draft Report, no. 445 (55/10/1), Ministry of Statistics and Programme Implementation, New Delhi.

Indian Council of Agricultural Research (ICAR), 1985, *Nutritional Requirement of Livestock and Poultry*, Report of the Scientific Panel on Animal Nutrition and Physiology, New Delhi.

Indian Council of Medical Research (ICMR), 1971, *Nutritive Value of Indian Foods*, New Delhi.

IPCC, 1995, *Guidelines for National Greenhouse Gas Inventories Reference Manual*, OECD, Paris.

Iype, Sosamma 1997, 'Vechur: Kerala's Dwarf Cow', *Dairy India.*

Jain, J.P., and Shivtar Singh, 1990, 'Qualitative and Quantitative Insufficiency of Livestock Feeds by 2001 AD and Possibilities of Bridging Gap', *Indian Journal of Animal Science*, vol. 60, no. 10, pp. 1224–9.

Jarvis, S.C., R.O. Lovell and R. Panayides, 1995, 'Patterns of Methane Emission from Excreta of Grazing Animals', *Soil. Biol. Biochem.*, vol. 27, no. 12, pp. 1581–8.

Johnson, K.A. and D.E. Johnson, 1995, 'Methane Emission from Cattle', *Journal of Animal Science*, vol. 73, pp. 2483–2.

Khadi and Village Industries Commission (KVIC), 1983, *Gobar Gas 'Why and How'*, Directorate of Gobar Gas Scheme, Bombay.

Krishna et al., 1978, 'Effect of Nutritional and Seasonal Variations on Heat and Methane Production in Bos Indicus', *Indian Journal of Animal Science*, vol. 48, no. 5, pp. 366–70, May.

Mishra, S.N., 1970, 'Some Inferences from Compositional Changes in India's Livestock Population 1920–66', *Indian Journal of Agricultural Economics*, vol. 25, no. 4, October–December.

——, 1978, *Livestock Planning in India*, Institute of Economic Growth, Vikas Publishing House, Delhi.

——, 1999, 'Dr. V.G. Panse Memorial Lecture: The Status of Livestock Statistics in India', *Journal of the Indian Society of Agricultural Statistics*, vol. 52, no. 3, pp. 273–89.

Mishra, S.N. and John Beyer, 1976, *Cost–Benefit Analysis: A Case Study of Ratnagiri Fisheries Project*, Institute of Economic Growth, Hindustan Publishing Corporation, Delhi.

Mishra, S.N. and R.K. Sharma, 1990, *Livestock Development in India: An Appraisal*, Institute of Economic Growth (IEG), Vikas Publishing House, Delhi.

National Bureau of Soil Survey and Land Use Planning (ICAR), 1992, *Agro-Ecological Region of India*, Technical Bulletin, NBSS Publication, no. 24, Nagpur, India.

Pandey, U.K., 1995, 'The Livestock Economy of India: A Profile', *Indian Journal of Agriculture Economics*, vol. 50, no. 3, July–September.

Patel, V.J., 1985, 'Rational Approach Towards Fuel Wood Crisis in Rural India', *Economic and Political Weekly*, vol. 20, no. 32.

Ranjhan, S.K., 1998, *Nutrient Requirements of Livestock and Poultry*, Indian Council of Agricultural Research, New Delhi.

Sen, K.C., S.N. Ray and S.K. Ranjhan, 1978, *Nutritive Values of Cattle Feeds and Feeding of Animals*, Indian Council of Agricultural Research, New Delhi (revised version of the ICAR Bulletin, no. 25 by K.C. Sen and S.N. Ray).

Sharma, R.K., 1987, 'Animal Energy in Agriculture: A Case Study of Uttar Pradesh,' Agricultural Economics Research Centre, University of Delhi, March.

Singh, G.P., 1997, 'Effect of Greenhouse Gases on Climate Change and Indian Ruminant Livestock', *Current Science*, vol. 72, no. 7, April.

Statistical Abstract of United States, 1995, 115th edition, The Reference Press, INC.

United States Department of Agriculture (USDA), Economic Research Service (ERS), 1991, *Corn: State-level Production Costs, Characteristics, Input and Machine Data*, Washington, D.C.

——, 1995, *Feed Outlook*, Washington, D.C., 13 January.

——, 1999, *Fertilizer Marketing News*, Washington, D.C., September.

United States Department of Agriculture (USDA), National Agriculture Statistics Service (NASS), *Agricultural Statistics, 1994 and 1997*, Washington, D.C.

United States Environmental Protection Agency (USEPA), *Inventory of U.S. Greenhouse Gas Emissions and Sinks, 1990–98*.

World Bank, 1999, *Indian Livestock Sector Review: Enhancing Growth and Development*, Rural Development Sector Unit, South Asia Region.